The publishing house tredition has created the series **TREDITION CLASSICS**. It contains classical literature works from over two thousand years. Most of these titles have been out of print and off the bookstore shelves for decades.

The book series is intended to preserve the cultural legacy and to promote the timeless works of classical literature. As a reader of a **TREDITION CLASSICS** book, the reader supports the mission to save many of the amazing works of world literature from oblivion.

The symbol of **TREDITION CLASSICS** is Johannes Gutenberg (1400 – 1468), the inventor of movable type printing.

With the series, tredition intends to make thousands of international literature classics available in printed format again – worldwide.

All books are available at book retailers worldwide in paperback and in hardcover. For more information please visit: www.tredition.com

tredition was established in 2006 by Sandra Latusseck and Soenke Schulz. Based in Hamburg, Germany, tredition offers publishing solutions to authors and publishing houses, combined with worldwide distribution of printed and digital book content. tredition is uniquely positioned to enable authors and publishing houses to create books on their own terms and without conventional manufacturing risks.

For more information please visit: www.tredition.com

Fruits of Queensland

Albert H. Benson

Imprint

This book is part of the TREDITION CLASSICS series.

Author: Albert H. Benson
Cover design: toepferschumann, Berlin (Germany)

Publisher: tredition GmbH, Hamburg (Germany)
ISBN: 978-3-8491-8655-5

www.tredition.com
www.tredition.de

Copyright:
The content of this book is sourced from the public domain.

The intention of the TREDITION CLASSICS series is to make world literature in the public domain available in printed format. Literary enthusiasts and organizations worldwide have scanned and digitally edited the original texts. tredition has subsequently formatted and redesigned the content into a modern reading layout. Therefore, we cannot guarantee the exact reproduction of the original format of a particular historic edition. Please also note that no modifications have been made to the spelling, therefore it may differ from the orthography used today.

Fruit of Mangosteen.

CONTENTS.

Preface

Introduction

Queensland Fruit-growing

Climate

1st. — Soils of Eastern Seaboard, and land adjacent to it, suitable to the growth of Tropical and Semi-tropical Fruit

2nd. — Soils of the Coastal Tablelands, suitable for the growth of Deciduous Fruit

3rd. — Soils of the Central Tablelands, suitable for the growth of Grapes, Dates, Citrus Fruits, &c.

The Banana

The Pineapple

The Mango

The Mangosteen

The Papaw

The Cocoa-nut

The Granadilla

The Passion Fruit

Custard Apples

Citrus Fruit

The Persimmon

The Loquat

The Date Palm

The Pecan Nut

Japanese Plums

Chickasaw Plums

Chinese Peaches

Figs

The Mulberry

The Strawberry
Cape Gooseberry
The Olive
The Apple
The Peach
The Plum
The Apricot
The Cherry
The Pear
The Almond
Grape Culture
List of Fruits and Vegetables Grown in Queensland

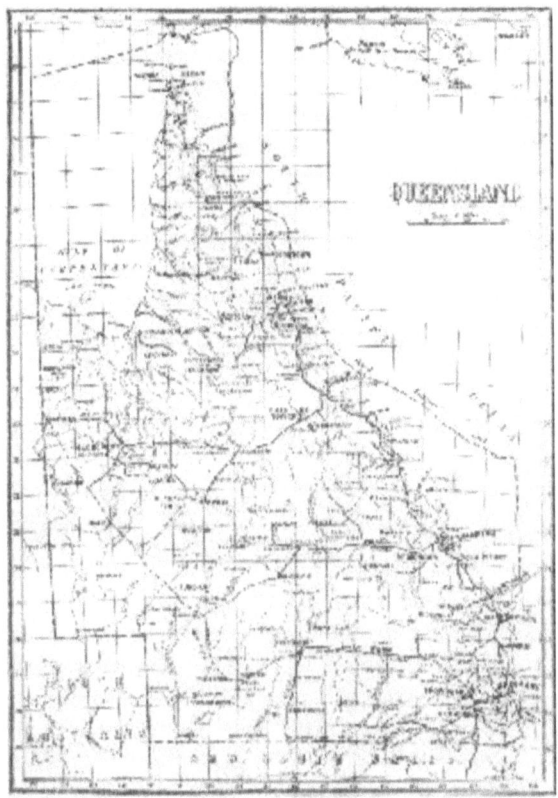

PREFACE.

In the more thickly populated portions of the Old and New World, and, to a certain extent, in the large cities of Australia, the question of how to make a living has became one of vital importance to a large portion of the population, and is the cause of considerable anxiety to fathers of families who are endeavouring to find employment for their sons.

This difficulty of obtaining employment is a very serious question, and one demanding the most earnest consideration. It is probably the result of many different causes, but, in the writer's opinion, it is due mainly to the fact that for years past the trend of population has been from the country districts to the towns, with the result that many of the great centres of population are now very badly congested, and profitable employment of any kind is often extremely difficult to obtain. The congested towns offer no possible outlet for surplus labour, hence it is necessary that such labour must find an outlet in the less thickly populated parts of the world where there is still plenty of room for development and population is badly needed. Queensland is a country possessing these qualifications; but is, unfortunately, a country that is little known to the general mass of home-seekers, and, further, what little is known of it is usually so inaccurate that a very erroneous opinion of the capabilities of this really fine country exists. The great flow of emigration is naturally to those countries that are nearest to the Old World—viz., the United States of America and Canada—and little attention is given to Australia, although we have many advantages not possessed by either the United States or Canada, and are not subject to the disadvantage of an intensely cold winter such as that experienced throughout the greater portion of those countries for several months yearly.

To those looking for homes the following pages are addressed, so that before deciding to what part of the world they will go they may know what sort of a country Queensland really is, what one of its industries is like, the kind of life they may look forward to spending here, and the possibility of their making a comfortable home amongst us. The life of a fruit-grower is by no means a hard one in Queensland, the climate of the fruit-growing districts is a healthy and by no means a trying one, and is thoroughly adapted to the successful cultivation of many fruits; and, finally, a living can be made under conditions that are much more conducive to the wellbeing of our race than those existing in the overcrowded centres of population. The writer has no wish to infer that there are big profits to be made by growing fruit, but, at the same time, he has no hesitation in saying that where the industry is conducted in an up-to-date manner, on business lines, a good living can be made, and that there is a good opening for many who are now badly in want of employment. The illustrations represent various phases of the industry, and have been specially prepared by H. W. Mobsby, the Artist of the Intelligence and Tourist Bureau. Most of the Illustrations have been taken at an exceptionally dry time, and at the close of one of the coldest winters on record, so that they do not show the crops or trees at their best; at the same time, they give a fair idea of some of our fruits, orchards, and fruit lands.

ALBERT H. BENSON.
Brisbane, Queensland, January, 1906.

INTRODUCTION.

Queensland's greatest want to-day is population: Men and women to develop our great natural resources, to go out into our country districts as farmers, dairymen, or fruit-growers—not to stick in our towns, but to become primary producers, workers, home-builders—not the scourings of big cities, the dissatisfied, the loafer, but the honest worker whose wish is to make a home for himself and his family. There are many such in the overcrowded cities of older countries, striving in vain to make a living—existing, it can hardly be called living, under conditions that are by no means conducive to their well-being—often poorly fed and poorly clad—who would better themselves by coming to Queensland, and by whom Queensland would be benefited. Queensland has room for many such annually: men and women who come here for the express intention of settling amongst us and building homes for themselves; who come here prepared to work, and, if needs be, to work hard; who do not expect to become rich suddenly, but will be contented with a comfortable home, a healthy life, and a moderate return for their labour—results that are within the reach of all, and which compare more than favourably with the conditions under which they are at present existing.

Queensland's most valuable asset is her soil, and this requires population to develop it: soil that, in the different districts and climates best adapted for their growth, is capable of producing most of the cultivated crops of the world, and, with very few exceptions, all the fruits of commercial value, many of them to a very high degree of perfection. This pamphlet is practically confined to the fruit-growing possibilities of Queensland, and an endeavour is made to show that there is a good opening for intending settlers in this branch of agriculture, but the general remarks respecting the climate, rainfall, soils, &c., will be of equal interest to any who wish to take up any other branch, such as general farming, dairying, &c. The Queensland Department of Agriculture has received a number of inquiries from time to time, and from various parts of the world, respecting the possibilities of profitable commercial fruit-growing in this State, and this pamphlet is intended in part to be an answer to such inquiries; but, at the same time, it is hoped that it will have a wider scope, and give a general idea of one of our staple industries to many who are now on the look-out for a country in which to settle and an occupation to take up when they arrive there.

Woombye, North Coast Railway. The centre of a large fruit growing district.

No branch of agriculture has made a greater advance during the past quarter-century than that of fruit-growing, and none has become more popular. The demand for fruit of all kinds, whether fresh or preserved, has increased enormously throughout the world, and it is now generally looked upon more as a necessity than a luxury. Hence there are continually recurring inquiries as to the best place to start fruit-growing with a reasonable prospect of success. It is not only the increased demand for fruit that causes these inquiries, but fruit-growing has a strong attraction for many would-be agriculturists as compared with general farming, dairying, or stock-raising, and this attraction is probably due to a certain fascination it possesses that only those who have been intimately acquainted with the industry for years can fully appreciate. In addition to the fact that living under one's own vine and fig-tree is in itself a very pleasant ideal to look forward to, there is no branch of agronomy that calls for a keener appreciation of the laws of Nature, that brings man into closer touch with Nature, that makes a greater demand on a man's patience, skill, and energy, or in which science and practice are more closely related, than in that of fruit-growing. To all those who are considering the advantages of taking up fruit-growing as an occupation, and to those who feel the attraction I have just described, these few words on fruit-growing in Queensland are addressed, as the writer wishes them to learn something of the fruit-growing capabilities of this State, so that before deciding on the country in which they will make a start they may not be in complete ignorance of a land that is especially adapted for the growth of a larger number of distinct varieties of fruit than any other similar area of land with which he is acquainted either in the Old or New World.

Queensland is a country whose capabilities are at present comparatively unknown even to those living in the Southern States of Australia, and, naturally, very much less so to the rest of the world, hence a little general information respecting our country and one of its industries may be of some help to those who are looking for an opening in this particular branch of agriculture.

Queensland is a country having a population of a little over half a million, and an area of 429,120,000 acres; the population of a city of the second magnitude, and an area of some seven and one-half times greater than that of Great Britain, or two and one-half times greater than the State of Texas, United States of America.

A Tropical Orchard, Port Douglas.

Coochin York Mangosteen, Port Douglas District.

A country embracing 18 degrees of latitude, from the 11th to the 29th degrees of south latitude, and extending from a humid eastern seaboard to an extremely dry interior, some 15 degrees of longitude west. A country, therefore, of many climates and varied rainfall. A country possessing a great diversity of soils, many of which are of surprising richness. A country more or less heavily timbered with either scrub or forest growth, or consisting of wide open plains that are practically treeless. A country of infinite resources, that is capable of producing within its own borders all that man requires, from the extreme tropical to temperate products. A country that, once its possibilities are realised and turned to a profitable account, is destined to become one of the most fruitful in the globe, to support a large and thriving population of our own people; and last, but not least, a country that, from a fruit-grower's point of view, cannot be excelled elsewhere. We have a healthy climate, not by any means an extreme climate as is often represented—extreme cold is unknown, frost being unusual on any portion of the seaboard, but common during the winter months on our tablelands. But even where there are frosts the days are pleasantly warm. Summer is undoubtedly warm, but it is usually a bearable heat, and sudden changes are extremely rare, so that though trying in the humid tropical seaboard, it is not unbearable, and compares favourably with the tropical heat met with elsewhere. This is clearly shown by the stamina of the white race, particularly those living in the country districts, where both men and women compare favourably with those of any other part of the Empire. Except in very isolated places, communication with the outside world and between the different centres of population is regular and frequent; in fact, in all the coastal and coastal tableland districts of the State one is kept daily in touch with all the important matters that are taking place in the world. In the home life there is a freedom not met with in older countries; there is an almost entire absence of artificiality—people are natural, and are interested in each other's welfare. They are certainly fond of pleasure, but at the same time are extremely generous and hospitable. The writer can speak of this from a large practical experience, as for some years past he has annually travelled many thousands of miles amongst fruit-growers and others who are settled on the land, and, without exception, he has everywhere been met with the greatest kindness from rich and poor alike—in short, a hearty welcome—and the best that the house affords is the rule, without exception. In brief, should any of my readers decide on coming to Queensland, the only difference that they will find as compared with the older countries is, that our climate is somewhat warmer in summer, but to compensate for this we have no severe cold in winter. There is more freedom and less conventionality, life to all who will work is much easier, and there is not the same necessity for expensive clothing or houses as exists in more rigorous climates. The people they will meet are of their own colour and race, no doubt fond of sport and pleasure, perhaps inclined to be a little self-opinionated, but solid grit at the bottom. As previously stated, Queensland offers exceptional advantages to the intending fruit-grower, and the following may be quoted as examples. The ease with which fruit can be produced, when grown under conditions suitable to its proper development, is often remarkable, and is a constant source of wonder to all who have been accustomed to the comparatively slow growth of many of our commoner varieties of fruits when grown in less favoured climes, and to the care that is there necessary to produce profitable returns. Here all kinds of tree life is rapid, and fruit trees come into bearing much sooner than they do in colder climates. In addition to their arriving at early maturity, they are also, as a rule, heavy bearers, their fault, if anything, being towards over-bearing. Fruits of many kinds are so thoroughly acclimatised that it is by no means uncommon to find them growing wild, and holding their own in the midst of rank indigenous vegetation, without receiving the slightest care or attention. In some cases where cultivated fruits have been allowed to become wild, they have become somewhat of a pest, and have kept down all other growths, so much so that it has been actually necessary to take steps to prevent them from becoming a nuisance, so readily do they grow, and so rapidly do they increase. The very ease with which fruit can be grown when planted under conditions of soil and climate favourable to its development has had a tendency to make growers somewhat careless as compared with those of other countries who have to grow fruit under conditions demanding the most careful attention in order to be made profitable. This is enough to show that Queensland is adapted for fruit-growing, and the illustrations accompanying the description of our chief commercial fruits will show them more forcibly than any words of mine that my contention is a correct one. Latterly, however, there has been a considerable improvement in the working of our orchards, growers finding that it does not pay to grow second-quality fruit, and, therefore, they are giving much more attention to the selection of varieties, cultivation of the land, pruning the trees, and the keeping in check of fruit pests; as, like other parts of the world, we have our pests to deal with. This improvement in the care and management of our orchards is resulting in a corresponding improvement in the quantity and quality of our output, so that now our commercial fruits—that is to say, the fruits grown in commercial quantities—compare favourably with the best types of similar fruits produced elsewhere. The writer has no wish to convey the impression that all that is required in order to grow fruit in Queensland is to secure suitable land, plant the trees, let Nature do the rest, and when they come into bearing simply gather and market the fruit. This has been done in the past, and may be done again under favourable conditions, but it is not the usual method adopted, nor is it to be recommended. Here, as elsewhere, the progressive fruit-growing of to-day has become practically a science, as the fruit-grower who wishes to keep abreast of the times depends largely on the practical application of scientific knowledge for the successful carrying on of his business. There is no branch of agronomy in which science and practice are more closely connected than in that of fruit-growing. Every operation of the fruit-grower is, or should be, carried out on scientific lines and by the best methods of propagation—pruning, cultivation, manuring, treatment of diseases, and preservation of fruit when grown are all, directly or indirectly, the result of scientific research. To be a successful fruit-grower in Queensland one must therefore use one's brains as well as one's hands; the right tree must be grown in the right kind of soil and under the right conditions; it must be properly attended to, and the fruit, when grown, must be marketed in the best possible condition, whether same be as fresh fruit or dried, canned, or otherwise preserved, and whether same be destined for our local, Australian, or oversea markets. Fruit-growing on these lines is a success in Queensland to-day, and it is capable of considerable extension, so that, in the writer's opinion, it offers a good field for the intending settler. Carried out in the manner indicated, he has no hesitation in saying that Queensland is a good place in which to start fruit-growing, that the advantages it possesses cannot be surpassed or even equalled elsewhere, and, further, that as our seasons are the opposite of those in countries situated on the north of the equator, our fruits ripen in the off-seasons of similar fruit grown in those countries, and, with our facilities for cold storage and rapid transit, can be placed on their markets at a time that they are bare of such fruits, thus securing top prices.

Bunch of Fruit of the Coochin York Mangosteen.

Queensland has practically an unlimited area of land suitable for fruit culture, much of which is at present in its virgin state, and is obtainable on easy terms and at a low rate. Government land is worth on an average £1 per acre, and privately-owned land suitable for fruit-growing can be purchased at from 10s. to £5 per acre, according to its quality and its distance from railway or water carriage. We have plenty of land, what we lack is population to work it; and there is no fear of over-crowding for many years to come. We have not only large areas of good fruit land at reasonable rates, but the Government of Queensland, through its Department of Agriculture, is always ready to give full information to intending settlers, to assist them in their selection of suitable land, to advise them as to the kinds of fruit to plant, to give practical advice in the cultivation, pruning, manuring, and general management of the orchard as well as in the disposal or utilisation of the fruit when grown; in short, to help the beginner to start on the right lines, so that he will be successful.

There is also little if any fear of over-extending the fruit-growing industry, as, if it is conducted on the right lines and on sound business principles, we can raise fruit of the highest quality at a price that will enable us to compete in the markets of the world especially now that we have direct and rapid communication at frequent intervals with Canada, the United States of America, the East (Japan, Manilla, &c.), Europe, and the United Kingdom.

Tamarind Fruits—Kamerunga State Nursery, Cairns.

QUEENSLAND FRUIT GROWING.

Very few persons have any idea of the magnitude or the resources of this State of Queensland, and in no branch of agricultural industry are they more clearly shown than in that of fruit-growing. Here, unlike the colder parts of the world or the extreme tropics, we are not confined to the growing of particular varieties of fruits, but, owing to our great extent of country, and its geographical distribution, we are able to produce practically all the cultivated fruits of the world, many of them to great perfection. There are, however, one or two tropical fruits that are exceptions, such as the durien and mangosteen, whose range is extremely small, and one or two of the berry fruits of cold countries, which require a colder winter than that experienced in any part of this State. It will, however, be seen at once that a country that can produce such fruits as the mango, pineapple, banana, papaw, granadilla, guava, custard apple, litchi, sour sop, cocoa nut, bread fruit, jack fruit, monstera, alligator pear, and others of a purely tropical character; the date, citrus fruits of all kinds, passion fruit, persimmon, olive, pecan nut, cape gooseberry, loquat, and other fruits of a semi-tropical character, as well as the fruits of the more temperate regions, such as the apple, pear, plum, peach, apricot, quince, almond, cherry, fig, walnut, strawberry, mulberry, and others of minor importance, in addition to grapes of all kinds, both for wine and table, and of both European and American origin, offers a very wide choice of fruits indeed to the prospective grower. Of course, it must not be thought for a moment that all the fruits mentioned can be grown to perfection at any one place in the State, as that would be an impossibility, but they can be grown in some part of the State profitably and to great perfection.

The law of successful fruit culture is the same here as in all other fruit-producing countries—viz., to grow in your district only those fruits which are particularly adapted to your soil and climate, and to let others grow those fruits which you cannot grow, but which their conditions allow them to produce to perfection. The intending grower must, therefore, first decide on what fruits he wishes to grow, and when he has done so, select the district best suited to their growth. The small map of the State shows the districts in which certain fruits may be grown profitably, or, rather, the districts in which they are at present being so grown; but there are many other districts in which fruit-growing has not been attempted in commercial quantities or for other than purely home consumption that, once the State begins to fill up with population, are equal, if not superior, to the older fruit-growing districts, and are capable of maintaining a large population.

Typical Clean Orchard.

CLIMATE.

As previously stated, the successful culture of fruit depends mainly on the right kinds of fruit being grown in the right soil and climate. This naturally brings us to the question of climate, and here one again gets an idea of the extent of our country, as we have not one but many climates. Climate is a matter of such vital importance to fruit-growers, and there is such a general lack of knowledge respecting the climate of Queensland, that a little information on this point is desirable. I am afraid that there is a very general impression that Queensland has a climate that is only suitable for a coloured race; that it is either in the condition of a burnt-up desert or is being flooded out. That it is a country of droughts and floods, a country of extremes—in fact, a very desirable place to live out of. No more erroneous idea was ever given credence to, and, as an Englishman born, who has had many years' practical experience on the land in England, Scotland, the United States of America, and the various Australian States, I have no hesitation in saying that, as far as my experience goes—and it is an experience gained by visiting nearly every part of the State that is suited for agricultural pursuits—taken as a whole, it is difficult to find a better or healthier climate in any other country of equal area. Our climate has its disadvantages, no doubt, particularly our dry spells, but show me the country that has a perfect climate. We have disadvantages, but, at the same time, we have great advantages; advantages that, in my opinion, outweigh our disadvantages.

Our eastern seaboard, extending from the New South Wales border in the south, a few miles to the south of the 28th degree of south latitude, to Cape York, some 20 miles north of the 11th degree of south latitude, contains our best districts for the growth of tropical and semi-tropical fruits. The coastal climate, however, varies considerably, and is governed by the proximity or otherwise of the coast ranges. When they approach the coast there is always more rainfall, and as they recede the rainfall decreases. With one or two exceptions, where the coastal range is a considerable distance inland, the eastern coastal districts have a sufficient rainfall for the successful culture of most fruits, though they are subject to a dry spell during winter and spring. During this period of the year, the weather is extremely enjoyable; in fact, it is hard to better it, even in our extreme North. But as summer approaches, thunderstorms become prevalent, and are accompanied by more or less humid conditions, which, though good for fruit-development, are not quite so enjoyable as the drier months. Summer is our rainy season, and the rainfalls are occasionally very heavy. The weather is warm and oppressive, particularly in the more tropical districts; but these very conditions are those that are best suited to the production of tropical fruits. The climate of those districts having the heaviest summer rainfall is somewhat trying to Europeans, particularly women, but it is by no means unhealthy, and in the hottest parts, having the coast range nearly on the coast, there is, within a few miles, a tableland of from 2,000 to 4,000 feet elevation, where the climate is cool and bracing, and where the jaded man or woman can soon throw off the feeling of lassitude brought about by the heat and humidity of the seaboard. In autumn the weather soon cools off, drier conditions supervene, and living again becomes a pleasure in one of the best and healthiest climates to be met with anywhere. Practically all the district under review has a sufficient rainfall for the growth of all fruits suitable to the climate, though there are occasionally dry spells during spring, when a judicious watering would be a great advantage. This does not imply a regular system of irrigation, but simply the conserving of surplus moisture in times of plenty by means of dams across small natural watercourses or gullies, by tanks where such do not occur, or from wells where an available supply of underground water may be obtained. The water so conserved will only be needed occasionally, but it is an insurance against any possible loss or damage that might accrue to the trees during a dry spell of extra length. So far, little has been done in coastal districts in conserving water for fruit-growing, the natural rainfall being considered by many to be ample; but, in the writer's opinion, it will be found to be a good investment, as it will be the means of securing regular crops instead of an occasional partial failure, due to lack of sufficient moisture during a critical period of the tree's growth. The average yearly rainfall in the eastern seaboard varies from 149 inches at Geraldton to 41 inches at Bowen, the mean average being about 90 inches to the north and 49 inches to the south of Townsville. Were this fall evenly distributed throughout the year, it would be ample for all requirements. Unfortunately, however, it is not evenly distributed, the heavy falls taking place during the summer months, so that there is often a dry spell of greater or less extent during the winter and spring, during which a judicious watering has a very beneficial effect on fruit trees, and secures a good crop for the coming season. The rainfall shows that there is no fear of a shortage of water at any time, the only question is to conserve the surplus for use during a prolonged dry spell. These conditions are extremely favourable for the growth of all tropical and semi-tropical fruits, as during our period of greater heat, when these fruits make their greatest call for moisture, there is an abundance of rain, and during the other portions of the year, when the call is not so heavy, it is usually an inexpensive matter to conserve or obtain a sufficient supply to keep the trees in the best of order. Throughout the southern half of this seaboard frosts are not unknown on low-lying ground, but are extremely rare on the actual coast, or at an elevation of 300 to 400 feet above the sea, so much so that no precautions are necessary to prevent damage from frost. We have, unlike Florida and other parts of the United States of America — great fruit-growing districts — no killing frosts, and now, at the close of one of the coldest winters on record, and one of the driest, nowhere have our pineapples — fruit nor plants — been injured, except on low-lying ground, over in the Southern part of the State, and mangoes, bananas, &c., are uninjured.

Burning-off for fruit growing, Mapleton, Blackall Range.

Same land one year later. Fruit-grower's family gathering strawberries.

In the more tropical North frosts are unknown on the coast, and there is no danger to even the most delicate plants from cold.

Running parallel with the coast we have a series of ranges of low mountains, running from 2,000 feet to nearly 6,000 feet, the general height being from 2,000 to 3,000 feet, and at the back of these ranges more or less level tablelands, sloping generally to the west. On and adjacent to these ranges in the Southern part of the State, there are fairly sharp frosts in winter, but the days are warm and bright. This is the district best adapted for the growth of deciduous fruits and vines, table varieties doing particularly well. It is a district well adapted for mixed farming and dairying, as well as fruit-growing; the climate is even and healthy, and is neither severe in summer nor winter. The average rainfall is some 30 inches, and is usually sufficient, though there are dry periods, when a judicious watering, as recommended for the coast districts, would be of great value to fruit and vegetable growers. The more northern end of this tableland country has a much better rainfall—some 40 inches per annum—and frosts, though they occur at times, are not common. Here the climate is very healthy, there are no extremes of heat and cold, and, lying as it does inland from the most trying portion of our tropical seaboard, it forms a natural sanatorium to this part of our State.

Further west the rainfall decreases, the summers are hot—a dry heat, as distinct from the more humid heat of the coast, and much more bearable. There are frequent frosts in winter, particularly in the Southern part of the State. Fruit-growing is only carried on to a slight extent at present, and then only with the help of water, but when the latter is obtainable, very good results are obtained. Grapes do well, both wine and table, and for raisin-making. Citrus fruits are remarkably fine, the lemons especially, being the best grown in the State. The trees are less liable to the attack of many pests, the dryness of the air retarding their development, if not altogether preventing their occurrence. The date palm is quite at home here, and when planted in deep sandy land, and supplied with sufficient water, it is a rapid grower and heavy bearer. As an offset to the smallness of the rainfall, there is a good supply of artesian water, distributed over a wide range of country, that can be obtained at a reasonable rate, and that is suitable for irrigation purposes. All bore water is not suitable for irrigation, however, as some of it is too highly mineralised, but there are large areas of country possessing an artesian supply of excellent quality for this purpose. It will thus be seen that we have in Queensland, roughly, three distinct belts of fruit-growing country—

1st.—The Eastern Seaboard, and the land adjacent to it, suitable for the growing of tropical and semi-tropical fruit;

2nd.—The Coastal Tablelands, suitable for the growth of deciduous fruits, vines, olives, and citrus fruits in parts;

3rd.—The Central Tablelands, suitable for the growth of grapes, for table and drying, dates, citrus fruits, &c., but requiring water for irrigation to produce profitably.

So far, I have confined my remarks mainly to the climatic side of fruit-growing, and, before dealing with the growing of the different kinds of fruit, I will say a few words about our fruit soils, and will deal with them in districts, as I have endeavoured to do in the case of climate.

1st.—Soils of Eastern Seaboard, and Land adjacent to it, suitable to the Growth of Tropical and Semi-Tropical Fruit.

Several distinct types of soil are found that are well adapted for fruit-growing, but they all have one general characteristic which is a *sine qua non* of success—viz., they must possess good natural drainage, so that there is no danger of their becoming waterlogged or soured during periods of continued or heavy rainfall, as these conditions are fatal to fruit culture under tropical and semi-tropical conditions. Of such soils, the first to be considered are those of basaltic origin. They are usually of a chocolate or rich red colour, are of great depth, in parts more or less covered with basaltic boulders, in others entirely free from stones. The surface soil is friable and easily worked, and the subsoil, which is usually of a rich red colour, is easily penetrated by the roots of trees and plants grown thereon. Occasionally the subsoil is more compact, in which case it is not so good for fruit-tree growth, but is better adapted for that of sugar-cane, corn, grass, &c. These basaltic soils are usually rich, and are covered in their virgin condition with what is termed scrub—a dense mass of vegetation closely resembling an Indian jungle. The scrub growth is totally distinct from forest growth, which will be described later, in that the bulk of the timber growing in it, much of which is of large size, is of a soft nature, and once cut down soon rots away. Imagine a dense wall of vegetation, consisting of large trees running up to 100 or 150 feet in height, with trunks ranging from 2 to 8 feet, or even more, in diameter, and between these trunks an impenetrable mass of smaller growths, all of the most vivid green colours, together with innumerable vines and creepers that are suspended from the branches of the trees, hanging in festoons, creeping palms and bamboos, ferns and orchids of many kinds, both on the ground and growing on the tree trunks, as well as many beautiful foliage plants only found in hothouses in England, and you will have a faint idea of what a virgin scrub in coastal Queensland is like. Much of the timber of the coastal scrubs is of considerable commercial value for building purposes and furniture making, and is, or should be, so utilised prior to felling and burning off.

True scrub lands are not by any means the most difficult to clear, though to a "new chum" the work will appear at first of a Herculean character. Brushing the dense undergrowth and then felling the timber at a face costs from £1 10s. to £2 per acre, according to density, size of timber, and proportion of hardwood trees contained in it, and once this is done the fallen mass is allowed to become thoroughly dry, when it is burnt off. A good fire is half the battle, as the subsequent work of burning off the heavy timber left from the first burn is comparatively light. No stumps are taken out, as the bulk are found to rot out in a few years, and their presence in the soil is no detriment to the planting of such crops as bananas or even citrus fruit trees. No special preparation of the land, such as breaking up, &c., is necessary prior to planting. Holes are dug, trees or bananas are planted, and the whole cultivation for the first few years consists in keeping down weed growths with the chipping hoe. Once the stumps have rotted out the plough and other implements of culture take the place of the hoe. These soils are especially adapted for the growth of oranges, limes, mandarins, mangoes, bananas, pineapples, papaws, custard apples, strawberries, and cape gooseberries in the South; in fact, for nearly every kind of tropical and semi-tropical fruit.

Some basaltic soils are occasionally covered with forest in the place of scrub, or a mixture, part scrub and part forest. Forest country, as distinct from scrub, is open-timbered country, with little undergrowth, and no vines or other creepers. The timbers are also, as a rule, very hard, and the stumps will not rot out. Such land, when at all heavily timbered, is much harder to clear and get ready for fruit-growing than true scrub, as all timber must be felled and burnt off, and all stumps and roots taken out, so that the land can be thoroughly broken up and brought into a good state of tilth prior to planting. These soils are suitable to the growth of similar fruits to the true scrubs, but, as a rule, they are not as rich. The second class of soils suitable to fruit-growing are of alluvial origin, and are of a sandy, loamy nature, of fair depth. They are usually met with along our creeks and rivers, or in the deltas of our rivers. In their virgin state they are either covered with scrub or forest, or a mixture of both, but the growth is seldom as strong as on the red volcanic soils. Heavy alluvial soils are not suitable for fruit culture, and are much more valuable for the growth of farm crops, but the light sandy loams and free loams of medium character suit all kinds of fruit to perfection. These soils usually are easy to work. They retain moisture well when well worked, and frequently they are capable of being irrigated, either from adjacent creeks or rivers, or by water from wells. These soils are some of our best for citrus fruits, and are well adapted for the growth of pineapples and bananas, as well as most other tropical fruits, when free from frosts. The third class of soils are free sandy loams, either scrub or forest. They are of various colours, and range in texture from light sandy loams to medium loams; they possess excellent drainage, and though, when covered with forest, they are not naturally rich, they make excellent fruit soils, and respond rapidly to systematic cultivation and manuring. They are usually of sandstone or granitic origin, and, when covered with scrub in the first place, grow good crops for the first few years, when they become more or less exhausted in one or more available plant foods, and require manuring. These soils, like the sandy alluvial loams, are easy to work, retain moisture well when kept in a state of perfect tilth, and respond readily to manuring. They will grow all kinds of fruits when free from frost. There are other soils on which fruit can be grown, but those mentioned represent those most suitable. The land on which these soils occur is often much broken, particularly in rich scrub country; it is fairly level when of alluvial origin, and more or less rolling, as a rule, when of a sandy loamy nature. High, ridgy, free, loamy country is usually the most free from frost, and alluvial flats the most liable to it.

2nd.—Soils of the Coastal Tablelands, suitable for the Growth of Deciduous Fruit.

Starting from the Southern part of the State, adjoining the New South Wales border, the fruit soils are all of granitic origin. The country is much broken, but between the ridges and along the creek flats there is a considerable area possessing soils varying from a coarse, granitic, gritty soil to a fine granitic soil; that on the creeks of an alluvial nature, but still granitic. These soils vary considerably in quality, but are, as a rule, easy to work and retain moisture well. They are covered with open forest and are particularly adapted to the growth of apples, plums, peaches, and grapes, though other deciduous fruits are grown but not to the same excellence as those mentioned. Proceeding north the fruit soils are either sandy loams or loams of a brownish colour of volcanic origin. The former are suitable for almonds and wine grapes, and the latter for peaches, apricots, pears, apples, and especially olives. Further north a few of these fruits may be grown on loamy soils, together with citrus fruits, but, commercially, deciduous fruits are confined to the southern end of this district, the winter temperature being too high for their successful growth further north, as the trees get no winter rest, hence do not mature their fruit-bearing wood properly.

3rd.—Soils of the Central Tablelands, suitable to the Growth of Grapes, Dates, Citrus Fruits, Etc.

At the Southern end of the State the fruit soils are all of a sandy nature. Nothing else is used in any quantity, as sandy soils alone will retain sufficient moisture for the growth of grapes and fruit trees during dry spells, and even then only when kept well and deeply worked. Further north, where suitable artesian water is available, the best fruit soils are also free loams of a sandy nature, either alluvial or open forest soils, but deep, and possessing perfect drainage, as irrigation on land without good natural drainage is fatal to fruit culture. These sandy loams are also easy to work; though by no means rich, they, on account of their depth, grow good crops of fruit by means of irrigation, and the fruit, such as dates, oranges, lemons, grapes, &c., is of very fine quality. The fruit soils of this district are covered either with open forest—the trees being of comparatively small size—or with a scrubby undergrowth through which a few larger trees are scattered. Nearly all the timber of this district is extremely hard, is more or less stunted, and burns readily, hence clearing is not a very expensive item.

Having now given a very brief description of our climate and the fruit-soils in our principal fruit-producing centres, we will next consider the culture of those fruits which are grown in commercial quantities in the different parts of the State, as well as that of a few less well-known fruits which show especial promise. We will first deal with our tropical fruits, of which the first to be considered is the banana, as its production greatly exceeds that of any other tropical fruit, and, as far as Australia is concerned, this is the only State in which it is grown in commercial quantities. From tropical fruits we will go on to semi-tropical fruits, then to temperate fruits and vines.

THE BANANA.

Under the heading of "Banana," all kinds of plantains will also be included, as they belong to one and the same family. The members of this family of plants are all tropical, and produce the most typical and best known tropical fruits.

The rank luxuriance of the growth of this class of fruits, their handsome foliage, their productiveness, their high economic value as food, and their universal distribution throughout the tropics, all combine to place them in a premier position. As a food it is unequalled amongst fruits, as no matter whether it is used green as a vegetable, ripe as a fruit, dried and ground into flour, or preserved in any other way, it is one of the most wholesome and nutritious of foods for human consumption. It is a staple article of diet in all tropical countries, and the stems of several varieties make an excellent food for all kinds of stock.

Cavendish Bananas on scrub land, Buderim Mountain.

Cavendish Bananas at Woombye on newly cleared land.

In Queensland, the culture of bananas is confined to the frostless belts of the eastern seaboard, as it is a plant that is extremely susceptible to cold, and is injured by the lightest frosts. It is grown in favourable locations in the South, where it produces excellent fruit, but its cultivation is much greater in the North, where the rainfall is heavier and the average annual temperature greater. In the Southern part of the State its cultivation is entirely in the hands of white growers, who have been growing it on suitable soil in suitable localities for the past fifty years or even more. I recently saw an old plantation that was set out over twenty years ago, and the present plants are still strong and healthy, and bearing good bunches of well-filled fruit, so that there is no question as to the suitability of the soil or climate. Bananas do best on rich scrub land, and it is no detriment to their growth if it is more or less covered with stones as long as there is sufficient soil to set the young plants. Shelter from heavy or cold winds is an advantage, and the plants thrive better under these conditions than when planted in more exposed positions. Bananas are frequently the first crop planted in newly burnt off scrub land, as they do not require any special preparation of such land, and the large amount of ash and partially burnt and decomposed vegetable mould provide an ample supply of food for the plants' use. Bananas are rank feeders, so that this abundance of available plant food causes a rapid growth, fine plants, and correspondingly large bunches of fruit. Though newly burnt off scrub land is the best for this fruit, it can be grown successfully in land that has been under cultivation for many years, provided that the land is rich enough naturally, or its fertility is maintained by judicious green and other manuring. In newly burnt off scrub land all that is necessary is, to dig holes 15 to 18 inches in diameter, and about 2 feet deep, set the young plants in it, and partly fill in the hole with good top soil. The young plant, which consists of a sucker taken from an older plant, will soon take root and grow rapidly under favourable conditions, producing its first bunch in from ten to twelve months after planting. At the same time that it is producing its first bunch it will send up two or more suckers at the base of the parent plant, and these in turn will bear fruit, and so on. After bearing, the stalk that has produced the bunch of fruit is cut down; if this is not done it will die down, as its work has been completed, and other suckers take its place. Too many suckers should not be allowed to grow or the plants will become too crowded, and be consequently stunted and produce small bunches. All the cultivation that is necessary is the keeping down of weed growth, and this, once the plants occupy the whole of the land, is not a hard matter. A plantation is at its best when about three years old, but remains profitable for six years or longer; in fact, there are many plantations still bearing good fruit that have been planted from twelve to twenty years. Small-growing or dwarf kinds, such as the Cavendish variety, are planted at from 12 to 16 feet apart each way, but large-growing bananas, such as the Sugar and Lady's Finger, require from 20 to 25 feet apart each way, as do the stronger-growing varieties of plantain. Plantains are not grown to any extent in Queensland, and our principal varieties are those already mentioned, the Cavendish variety greatly predominating. In the North, the cultivation of this latter variety is carried out on an extensive scale, principally by Chinese gardeners, who send the bulk of their produce to the Southern States of the Commonwealth. The industry supports a large number of persons other than the actual producers of the fruit, and forms one of our principal articles of export from the North. As many as 20,000 or more large bunches of bananas frequently leave by a single steamer for the South, and the bringing of this quantity to the port of shipment gives employment to a number of men on tram lines and small coastal steamers. The shipment of a heavy cargo of bananas presents a very busy scene that is not soon forgotten, the thousands of bunches of fruit that are either piled up on the wharf or that are being unloaded from railway trucks, small steamers or sometimes Chinese junks, forming such a mass of fruit that one often wonders how it is possible to consume it all before it becomes over-ripe. Still, it is consumed, or, at any rate, the greater portion of it is, as it is the universal fruit of the less wealthy portion of the community, the price at which it can be sold being so low that it is within the reach of everyone. A banana garden in full bearing is a very pretty sight, the thousands of plants, each with their one or more bunches of fruit, as, where there are several stems it is not at all uncommon to find two or more bunches of fruit in different states of development on the same plant, forming a mass of vegetation that must be seen to be appreciated. This is the case even with dwarf-growing kinds, but with strong-growing varieties, such as the Lady's Finger, the growth is so excessive that the wonder is, how the soil can support it.

Twenty-dozen Bunch, Buderim Mountain.

Bananas do remarkably well in Queensland, and there is practically an unlimited area of country suitable for their culture, much of which is at present in a state of Nature. Only the more easily accessible lands have been worked and of these only the richest. Manuring is unknown in most parts, and as soon as the plantation shows signs of deterioration it is abandoned, and a fresh one planted out in new land, the land previously under crop with bananas being either planted in sugar-cane or allowed to run to grass. This is certainly a very wasteful method of utilising our land, and the time will come, sooner or later, when greater care will have to be given to it, and that once land has become impoverished by banana culture, it will have to be put under a suitable rotation of crops, so as to fit it for being again planted to bananas. The trouble is, as I have already stated, we have too much land and too few people to work it, hence, so far, we are unable to use it to anything like the best advantage. During the year 1904 the production of bananas in Queensland was some 2,000,000 bunches, and when it is considered that each bunch will average about 12 dozen fruit, it will be seen that already we are producing a very large quantity. There is, however, plenty of room for extension, and any quantity of available country, but before this extension can be profitable, steps will have to be taken to utilise the fruit in a manner other than its consumption as fresh fruit, and this in itself will mean the opening up of new industries and the employment of a considerable amount of labour. I have mentioned 12 dozen as being the average quantity of fruit per bunch, but it is frequently much more than this, and I have often seen bunches of 25 to 30 dozen fine fruit grown on strong young plants on rich new land. Although the industry in the North is now almost entirely in the hands of Chinese gardeners, there is no reason whatever why it should not be run by white growers, as is done in the South, and there is no question that our white-grown bananas in the South compare more than favourably with the Northern Chinese-grown article, despite the fact that the latter has every advantage in climate and an abundance of virgin soil. Most of the photos of bananas are, I am sorry to say, not by any means typical of this industry, as they have been taken during the off-season, when the plants look ragged and are showing little new growth, and the bunches also are much smaller than usual. Still, I hope that the illustrations will give some idea of the growing and handling of this crop, and will show what a banana plant and its bunch are like.

Bananas for shipment at Innisfail.

THE PINEAPPLE.

If there is one fruit that Queensland can grow to perfection, it is undoubtedly the pineapple. This is not merely my own personal opinion, but is the universal admission of all who are qualified to judge. On many occasions I have taken men thoroughly conversant with pineapple-growing, and who knew what a good fruit really is, through some of our plantations, where I have given them fruit to test, and, without exception, they have had no hesitation in saying that they have never tasted better fruit. Our fruit has a firmness, freedom from fibre, and a flavour that is hard to beat. It is an excellent canning fruit, superior in this respect to the Singapore article, which it surpasses in flavour. This is admitted by English and European buyers, and its superiority is bound eventually to result in a great increase in canning and the establishment of large works run on thoroughly up-to-date lines.

Picking Pines for market—Woombye District.

Pineapple Plantation — showing plants of different ages —
Woombye, North Coast Line.

Like the banana, the pineapple is a tropical fruit, and is very sensitive to cold, hence its culture is confined to frostless districts. It is grown all along our eastern seaboard, where, when planted in suitable soils and under suitable conditions, it is, undoubtedly, our hardiest fruit, and is practically immune from any serious disease. Its culture is entirely in the open, no shelter whatever being given, so that we are not put to the great expense that growers of this fruit in Florida and some other pineapple-producing countries must incur if they wish to secure a crop. Here we have no severe freeze-outs, and, though dry spells retard the growth at times, we have never suffered any serious injury from this cause. In the Southern part of the State, the coolness of the winter retards growth somewhat, and occasionally the tops of the leaves and young fruit are slightly injured, particularly in low-lying land, or where the plants are growing on land having a cold subsoil. When grown under more favourable conditions, however, they sustain no injury, and produce fruit, more or less, all the year round. Pines are always in season, though there are times when they are comparatively scarce. There are usually two main crops a year — viz., a summer and a winter crop. The former is the heavier of the two, and the fruit is decidedly the best, as its sugar contents are much higher. The main summer crop ripens in the North from the beginning of November, and in the South from January to as late as March in some seasons. The main winter crop is usually at its best in July and August, but there is always more or less fruit during the other months of the year. The pineapple likes a warm, free, well-drained soil, that is free from frost in winter, and that will not become soured by heavy rain during summer. Sandy loams are, therefore, our best pineapple soils, though it does well on free loams of basaltic or alluvial origin. Unlike the banana, the pineapple does not do too well in newly burnt off scrub land, owing to the difficulty in working the ground and keeping it clean. It requires a thorough preparation of the soil prior to planting in order to be grown to perfection. In the case of new land of suitable texture, the timber should all be burnt off, and all stumps and roots taken out of the soil, which should then be carefully broken up and reduced to a fine tilth, all weed or grass growth being destroyed. It should then be again ploughed, and, if possible, subsoiled, so as to permit of the roots penetrating the ground to a fair depth instead of their merely depending on the few top inches of surface soil. Careful preparation of the land and deep stirring prior to planting will be found to pay well, and turn out far the cheapest in the end. Given suitable soil, well prepared, the growing of pineapples is not at all difficult, as the plants soon take root, and once they became established, they prove themselves to be extremely hardy. Pines will grow and thrive on comparatively poor soil, provided it is of suitable texture, but in such soils it is necessary to supplement the plant food in the soil by the addition of manures, if large fruit and heavy crops are to be obtained. Pineapples are propagated by means of suckers coming from the base of fruit-bearing plants, or from smaller suckers, or, as they are termed, robbers or gill sprouts that start from the fruiting stem just at the base of the fruit. They are also sometimes propagated by means of the crown, but this method is usually considered too slow. Well-developed suckers are usually preferred, as these come into bearing earliest, but equally good, if not better, returns are obtained by planting gill sprouts. The latter have the advantage in that they always develop a good root system before showing signs of fruit, hence their first crop is always a good one, and the fruit is of the best, whereas suckers sometimes start flowering as soon as they are planted, before they are properly established, with the result that the first fruit is small and inferior, and the plants have to throw out fresh suckers before a good crop is produced. Gill sprouts are slower in coming into bearing than suckers, but the results are usually more satisfactory. Like the banana, once a pineapple plant has borne fruit the fruiting stalk dies down, and its place is taken by one or more suckers, which in their turn bear fruit and die. Pineapples are planted in Queensland in several ways, but by far the most common method is to set the suckers out in single or double rows, from 8 to 9 feet apart, with the plants at from 1 to 2 feet apart in the row. The rows soon increase in width by the growth of suckers, and the throwing up of ratoons—surface roots thrown off from the original plant, which send up plants from below the ground as distinct from suckers, which come from the base or even higher up the stem of a fruiting plant. It is not at all an uncommon thing to see the rows grown together, so that the plantation appears to be a solid mass of plants, but pathways have to be kept between the rows to permit of gathering the fruit, manuring, &c. Pineapples have been grown in the Brisbane district for the past sixty years, and I have been shown beds of plants that have not been replanted for over forty years that are still producing good fruit. This shows how well at home this fruit is with us; but, in my opinion, it is not desirable to keep the plants so long in the same ground, as the finest fruit is always obtained from comparatively young plantations, the older ones producing too large a proportion of small fruit. From the Brisbane district this fruit has spread all over the eastern coast, and its production is increasing rapidly in several districts. Once the pine is planted, its cultivation is comparatively simple. If in single or double rows, all weed growth is kept down between the plants, and the ground between the rows is kept in a state of good cultivation by means of ploughing or cultivating, the soil being worked towards the rows so as to encourage the formation of suckers low down on the fruiting plants. Manure is given when necessary, the manure being worked in on either side of the rows.

Smooth-leaved Cayenne Pines in fruit, planted 15 months, Woombye District.

The pineapple comes into bearing early, and, except where suckers throw fruit as soon as planted, bear their first crop in from twelve to twenty months, according to the type of suckers planted and the time of year at which they are set. Practically every sucker will produce a fruit at the first fruiting, and these will be followed by succeeding crops, borne on the successive crops of suckers, so that when the whole of the ground is occupied by plants, the returns are very heavy. One thousand dozen marketable fruits is by no means an unusual crop for Queen pines in a plantation in full bearing, and, taking these at an average of 2½ lb. each, you get a return of 30,000 lb., or 15 tons American per acre. The illustrations herewith give a good general idea of the usual method of growing pines, and the method of handling and marketing, as well as of the nature of the country on which they are grown. The illustrations are mostly of smooth-leaved pines, which bear a fruit averaging from 6 to 8 lb. each, but occasionally running up to as much as 14 to 16 lb., though the latter is an extreme weight. The single pine shown is just under 12 lb. Several kinds of pines are grown, which are generally classified into roughs and smooths. The rough, or rough-leaved pines, such as the Common Queen and Ripley Queen, and local seedlings raised from them, are very prolific, and though not equal in size and appearance to the smooth-leaved Cayenne, our principal smooth-leaved kind, are usually considered to be of superior flavour, and to be better for canning or preserving. Rough pines run up to as much as 6 lb. weight each, but this is uncommon, the best average I have met with being about 4 lb. per pine, and they were exceptionally good. The price at which this fruit sells here seems absurd to those living in cold countries, who are accustomed to look upon it as a luxury only found on the tables of the wealthy, as good rough-leaved pines are worth about 1s. per dozen during the summer season, and smooth-leaved pines from 1s. 6d. to 2s. 6d. a dozen. Prices are certainly higher during the off-season, but growers would be well satisfied to get 1s. per dozen for rough pines all the year round. I have no hesitation in saying that pines can be grown at a profit at from £3 to £4 per ton, so that the cost of growing is so low that there is nothing to prevent us from canning the fruit and selling it at a price that will defy competition.

Pineapple Plantation—Pines packed for market, and showing fruit-grower's home, Woombye District.

Pineapple-growing has been a very profitable industry, particularly in the older plantations of the Brisbane district, and still continues to be so in many places despite the fact that prices are much lower now than they were some years since. The plantations from which the illustrations are taken are comparatively new ones, the land having been in its virgin state from six to eight years ago, and, as shown, some is only now being cleared. The owners of the plantations started without capital, and, by dint of hard work and perseverance, are now reaping an excellent return of some £50 per acre net profit. This is by no means an isolated example, but is one that is typical of what can be done, and has therefore been chosen. There is a great opening for the culture of this fruit in Queensland, and its cultivation is capable of being extended to a practically unlimited extent. We have a large amount of land suitable for the growth of this fruit that is available in different parts of the State, much of it at very reasonable rates, so that there is no difficulty in this direction for anyone wishing to make a start. It is an industry from which returns are quickly obtained, and is a branch of fruit-growing that holds out strong inducements and every prospect of success to intending growers. At present our production is about sufficient for our presently existing markets, but there is nothing to prevent these markets being widely extended. Our present means of utilising our surplus fruits, by canning or otherwise preserving same, are by no means as complete or up to date as they should be, and before they can become so, it is necessary to greatly increase our output. Small works cost too much to run as compared with large canning establishments, hence we are not yet in a position to make the most of our fruit. With increased production we will have an increase in the facilities for utilising the fruit. This requires labour, and there is right here an opening for many industrious workers, a business that I have no doubt will pay from the start, a business of which we have the Australian monopoly, and in which there is no reason that I can see in which we should not compete satisfactorily in the markets of the world.

Pineapple Plantation—Showing method of growing the fruit, Woombye District.

Queensland possesses many advantages respecting the growth of this fruit as compared with other countries in which it is grown commercially, which may be briefly enumerated as follows:—

1st.—Freedom from loss by freeze-outs;

2nd.—The ease with which the fruit can be grown, and its freedom from disease;

3rd.—The large area of land suitable to its culture, and the low price at which suitable land can be obtained;

4th.—The fine quality of the fruit;

5th.—The superiority of our fruit for canning purposes;

6th.—The low price at which it can be produced, and the heavy crops that can be grown.

These are enough reasons to show that in the pineapple we have a fruit well suited to our soil and climate, a fruit in the cultivation of which there is room for great extension, and which will provide a living for many industrious settlers.

Rough-leaved Pines, Redland Bay District.

Pineapple Plantation—On virgin soil, showing scrub land at back being cleared for fruit growing, Woombye District.

THE MANGO.

This magnificent fruit, which is practically unknown outside of the tropics, has become as hardy as a forest tree throughout our eastern seaboard, wherever it is planted out of frost. It has been named, and well named too, the apple of Queensland, as it stands as much neglect, and can be grown with as little care and attention as, or even less, than that given to the apple-trees in many of the Somerset or Devonshire orchards. It will not, however, stand frost. Droughts and floods have little effect on it; it will grow in any soil, from a sand to a heavy loam, amongst rocks, or on a gravelly or shaley land. Naturally, it does best in good land, but there are hundreds of cases where trees are doing well and bearing heavily on land that is by no means fruit land. The mango is one of our handsomest fruit trees; the symmetry of its growth, its large glossy leaves, the delicate colouring of its young growth, which is of different shades in different varieties, the abundance of fruit that it produces, varying in colour from dull-green to yellow, red, or even purplish tints, all render it conspicuous. As well as being one of our handsomest, it is also one of our most widely distributed fruits, being found growing luxuriantly the whole length of our eastern seaboard. A few trees are also to be met with inland in districts that are free from frosts, so that it stands both the dry heat of the interior and the humid heat of the coast. As a tropical fruit it naturally reaches its greatest perfection under our most tropical conditions, the trees there growing practically wild, requiring little if any attention, making a rapid growth, coming into bearing early, and producing heavy crops of fruit. Further south the growth is somewhat slower, though the trees grow to a large size and bear heavily. It is one of the easiest of trees to grow, as it is readily propagated by means of seed. In many plantations thousands of young seedlings may often be seen growing under the old trees, the seeds having taken root without even having been planted. In most cases it is propagated from seed, the stones of fruit showing especial merit being planted either in a nursery, or, better, still, where the tree is to remain permanently, as it usually does better when so planted than when grown in a nursery and thence transplanted to its permanent location. The land should be well worked prior to planting, and the young trees require to be kept free from weeds and undergrowth till such time as they occupy the whole of the ground, when they are able to look after themselves, and require no further attention, at any rate in the warmer parts. It is not at all uncommon to come across a mango-tree, in full bearing, in vigorous health, that is growing wild, the result of a stone that has been thrown away by someone who has eaten the fruit. The young tree has not only been able to hold its own against all kinds of indigenous growths, but has developed into a vigorous, healthy tree, thus showing that it is perfectly at home, and that the soil and climate of Queensland suit it to perfection. The fact that by far the greater portion of our mango-trees have been grown from seed has resulted in the production of innumerable varieties, many of which are of decidedly inferior quality, as one never knows when planting the seed what the resultant fruit is going to be like. One is more likely to get good fruit by planting the seeds from selected fruit of the highest quality, but is by no means certain to do so, as a number of seeds always revert to inferior types. This has had a bad effect on our mango industry, and has been apt to give the fruit as a class a bad name, so much so that we find it difficult to get our Southern neighbours to take to it at all readily. I can quite understand anyone, whose first experience of a mango is that of an inferior fruit, full of fibre, and having a distinctly disagreeable flavour, condemning the particular fruit, but because there are inferior fruits one should not condemn the whole without knowing what a really good mango is like.

Mango Trees, Port Douglas.

We have many good mangoes in Queensland, but only a few that are really first-class, and of the latter I have yet to meet the man or woman, who is a fruit-eater, who does not appreciate their exquisite flavour, and who does not consider them worthy to rank with any of the finest fruits. By many a really fine mango is considered to be the king of fruits, and I am not at all certain that they are not right, but, at the same time, a really bad mango is indescribably bad.

The mango grows to a large size here, even when comparatively young. I know trees over 50 feet in height, having a spread of the branches of more than 60 feet, a main trunk nearly 3 feet in diameter, that are under thirty years old, and that have borne from 1 to 2 tons of fruit for a single crop. Hundreds of tons of fruit go to waste annually for want of a market, or are consumed by farm animals, as the consumption of the fruit is practically confined to this State, and the production is greater than we can consume, despite the fact that mangoes are in season from the end of September to March, and that they are a favourite fruit with all who have acquired a liking for them. In addition to the consumption of the fruit in its fresh state, a quantity is converted into chutney, but this is so small that it has no appreciable effect on the crop as a whole. The unripe fruit makes an excellent substitute for apples, and is used stewed or for pies or tarts, and when sliced and dried it may be stored and used in a similar manner to dried apples.

Mango Tree near Brisbane.

In addition to its value as a fruit, the mango forms a handsome ornamental tree, and one that provides a good shade for stock. It is very free from disease, as with the exception of one or two species of scale insects, which do not cause any very serious damage, it has few serious pests. It is a fruit that is bound sooner or later to come into more general favour, particularly when the qualities of the finer varieties are better known. Until quite recently it was considered to be one of the most difficult trees to propagate by means of grafting or budding, hence its propagation has been practically confined to raising it from seed, but now we have found out how to work it by means of plate-budding, and are able to perpetuate our best sorts true to kind. This is sure to lead to a general improvement of our existing varieties, as old trees can be worked over by this means, or young trees of approved kinds can be grown in a nursery and distributed.

The fruit is very wholesome, is much appreciated by all who have acquired a taste for it, can be used fresh or dry, ripe or unripe, and cans well. It is a great addition to our list of purely tropical fruits, and finds a place in all orchards or gardens where it is capable of being grown.

THE MANGOSTEEN.

Many attempts have been made during past years to introduce this delicious fruit into Queensland, but these always resulted in failure. True, a certain variety of mangosteen has been successfully grown at Port Douglas, also on the Lower Burdekin, and rumours of the existence of the true Java mangosteen (*Garcinia mangostana*) have been received, but, in nearly every case, they have, on investigation, proved to be *Garcinia xanthochymus*, or some other species. At the Kamerunga State Nursery, however, trees of undoubted parentage were successfully raised. It is said that a thriving young plant, which is unquestionably *G. mangostana*, is owned by Mr. Banfield, of Dunk Island. The records of the Kamerunga Nursery show that in October, 1891, a quantity—about 100—of ripe mangosteen fruit was received from the Batavian agency by the then manager, Mr. Ebenezer Cowley, from which some 600 seeds were obtained. Of these, only a few germinated. The next mention is of the distribution, in February, 1892, of six plants to an applicant on the Mossman, and of two more in May of that year. Since then several young trees have been raised at the nursery, and one of them, in January, 1913, fruited for the first time for twenty-two years, and is the first to have done so in this State. Some of the fruit was sent to the Department of Agriculture and Stock, and proved to be fully equal to those of Java. A full history of the mangosteen and of its introduction into Queensland is given in "The Queensland Agricultural Journal" (vol. xxx., June and July, 1913). The photographs were taken from the original fruit.

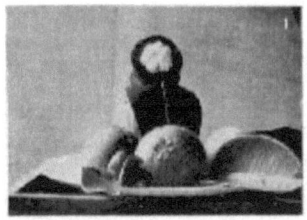

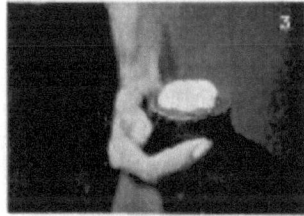

Fruit of Mangosteen.

THE PAPAW.

Continuing our list of tropical fruits, we now come to the papaw, one of our most wholesome and useful fruits. It is grown all along our eastern seaboard in situations that are free from frost. It comes into bearing early, and is a heavy cropper. Like the other tropical fruits already described, it does best in our warmer parts, coming to maturity earlier, and producing better fruit. In many of the Northern coastal scrubs it is often met with growing wild, and producing fruit in abundance, the seeds from which the trees have been produced having been dropped by birds or distributed by other natural agencies. The papaw fruit resembles a rock melon somewhat in shape and flavour, the fruit being produced in the axil of the leaves all along the main stem, where they are clustered thickly together. The tree does best on well-drained soils, and is very sensitive to the presence of clay or stagnant water at the roots, hence it usually does best on scrub land or land well supplied with humus. It is propagated entirely from seed, which grows readily in such soils, and under favourable conditions will bear its first fruit when about ten to twelve months old, and continue to bear for three or four years or even longer. When the trees becomes old, however, the fruit decreases in size and deteriorates in quality, so that it is necessary to plant a number yearly in order to keep up a regular supply. It is a very handsome tree, with large spreading leaves on long stems, beneath which is its cluster of fruit—as many as 100 fruits being sometimes found in different stages of development on the one plant. The fruit ranges in size from 2 lb. to some 6 lb. in weight, and when ripe it is of a greenish-yellow or sometimes orange colour. The flesh is yellow, and when quite ripe it is moderately juicy, and of a flavour that it not always appreciated at first, but which one soon becomes very partial to. It more nearly resembles the flavour of a rock melon than that of any other fruit, and the seeds, which are found clustered in the centre of the fruit, have a flavour that closely resembles that of seeds of the nasturtium. Both the seeds and the fruit contain an active principle called papain, which is really a vegetable pepsin, that has the effect of greatly assisting in the assimilation of all food with which it is eaten, hence it is a valuable remedy in the case of dyspepsia, and persons who take the fruit regularly are never subject to this exceedingly troublesome disease. The fruit can be used both as a vegetable and as a fruit, the former in its green state, when it is boiled and served with melted butter, resembles a vegetable marrow or squash, but is superior to either of these vegetables. As a fruit it is either used by itself, or in conjunction with other fruits it forms the basis of a fruit salad. It is largely used in the North, and its cultivation is steadily spreading South, as its valuable properties are becoming better known. Its cultivation is very simple. The seeds are either planted where the tree is to remain, or are raised in a bed and transplanted to their permanent position in the orchard when strong enough to stand shifting, care being taken to select a dull moist day. The young plants are protected from the sun for a few days till they have become established, after which all that is necessary is to keep down weeds and to work the soil round them, taking care not to injure the roots. A good mulch of decomposed vegetable matter round the plants is an advantage, but they are usually so easily grown that little extra care is given to them. The papaw bears male and female flowers, which may be on the same trees, but are usually on different trees, so that it is usual to speak of male and female trees. This is, however, a mistake, as according to Bailey the plant is polygamous—that is to say, male, female, or hermaphrodite flowers may be found on the same or on distinct plants. The male flowers are usually on long scantily-branched auxiliary panicles, whereas the female flowers are mostly in the axils of the leaves close to the stem. The two trees are not distinguishable from each other till they come into flower, hence it is advisable to set the young plants fairly close together—say, 6 feet apart—and thin out the male trees when same can be distinguished by their blossoms.

Besides its use as a fruit and vegetable, the papaw makes a fair conserve and an excellent sauce, and its medicinal principle, "papain," is an article of commerce.

Papaw in fruit, near Brisbane.

THE COCOA-NUT.

Although this palm can be grown for ornamental purposes as far south as Brisbane, its cultivation on commercial lines will be confined to the coast district north of Townsville, and to the islands off the coast, as, in order to develop its fruit to perfection, it requires a tropical climate. Where the climate is suitable it does well, it makes a rapid growth, and bears heavy crops of nuts. Old palms on the beach at Cairns compare favourably with any growing in the South Seas, and I am of opinion that its culture in commercial quantities on suitable land will be found profitable. The cocoa-nut palm does best right on or adjacent to the seashore, in comparatively poor sandy soil—soil that is usually of little value for general crops, though it will grow mangoes well. So far, it is not grown in any large numbers, and although there is a ready sale for the ripe nuts, there is no attempt to make copra or to utilise the coir. Copra is the dried flesh of the nut, from which oil is extracted, and is largely used in the manufacture of soap, candles, &c., the refuse left after the oil has been extracted being used for cattle feed. Coir is the fibre surrounding the nut, and is used for the manufacture of matting, door mats, &c.

There is a considerable area of land suitable to the culture of this fruit on our Northern coast, which is at present lying idle, that, in my opinion, can be turned to a profitable use by planting it in cocoa-nuts as, in addition to utilising land otherwise of little value, we would be building up a new industry. The trees come into bearing in about eight years after planting the seed, and will continue to produce crops for many years without any attention. Care will have to be given for the first few years, whilst the plants are small, to keep down undergrowth and to prevent fires from running through the plantation, but, once fairly established, the plants will look after themselves. A cocoa-nut plantation gives a distinctly tropical look to the district in which it is grown, and the palms, particularly when young, are very ornamental; when old the long bare stems detract somewhat from the beauty of the top. It is a palm that I believe has a good future before it in the North, and for that reason I have included it amongst our tropical fruits, though it is cultivated at present more as an ornamental plant than as an article of commerce.

Cocoa-nut Palms, Port Douglas.

THE GRANADILLA.

A vine, belonging to the natural order Passifloreæ, that produces one of our most delicious tropical fruits. The papaw and the passion fruit belong to this same order. It can be grown all along our eastern seaboard, but comes to greatest perfection in the North. The fruit is of a pale greenish-yellow colour, cylindrical in shape, and varies in weight from about 1 to 5 lb., the largest fruits being produced on a sub-species. The fruit consists of an outer pulpy covering, which can be used for cooking if desired, which surrounds a cavity filled with seeds which are encased in a jelly-like mass. This is the portion eaten, and to use an Americanism, "It is not at all hard to take." It is either eaten by itself, or is used in conjunction with papaw and other fruits to make a fruit salad, a dish that is fit for the food of the gods, and once taken is never forgotten.

The granadilla is easily grown from seed, and the plants are trained on an overhead trellis, the fruit hanging down on the underside. It is a heavy bearer, and once planted requires little attention. It requires a free, warm soil, that is fairly rich, to be grown to perfection, hence it is most commonly grown on scrub land. It can, however, be grown on any well-prepared land of a free nature. Unfortunately, it is a difficult fruit to ship any distance, hence its consumption is mainly confined to the districts in which it is grown, and where, needless to say, it is greatly appreciated. It is in fruit more or less all the year round, its main crop being in early spring in the North, and during the summer months further South. It is sometimes made into jam or jelly, but when preserved loses much of its characteristic flavour.

Granadilla Vine at Kuranda, Cairns district.

THE PASSION FRUIT.

This fruit is very closely related to the granadilla, but is much hardier than it, and can be grown to perfection much further South. It is not injured by frost to any extent in any part of coastal Queensland, and can be grown a considerable distance inland. It is more rightly a semi-tropical than a tropical fruit, though, as it is so nearly related to the granadilla, I have included it amongst the tropical fruits. It is also a vine, and, when grown commercially, is trained along a horizontal trellis, in a somewhat similar manner to a grape vine. It is readily grown from seed, and will produce fruit in less than twelve months from the time that it is planted, and will continue to bear fruit for some years. It does best on a free, warm soil of fair quality, though it may be grown anywhere with care, and often thrives well in very poor soils with the addition of manure. It is found growing wild on the borders of many of our scrubs and elsewhere, the seeds having been deposited by birds or other agencies, and under such conditions it produces an abundance of fruit. The fruit is of a roundish oval shape, and is of a dark-purple colour. It is about the size of a large hen's egg, the outer skin being hard and shell-like, and the centre filled with the seeds, which are surrounded with a jelly-like mass and a yellowish pulp. It is a very fine flavoured fruit, and is universally liked. It is grown in considerable quantities in the Southern part of the State, and is one of our commonest fruits. It has usually two crops a year—a summer and a winter crop—but can be got to produce its fruit at any particular time that is desired by systematic pruning at different times of the year. It is often grown over sheds, dead trees, fallen logs, &c., which it covers with a mass of dense green foliage, and converts what would otherwise be an unsightly object into an ornament. The illustration herewith shows this well, and gives a good idea of the growth of a single vine. Commercially it is grown on trellis, so that the land between the rows can be kept well cultivated, and also to permit of ease in the gathering of the fruit. When ripe, the fruit drops, and the gathering is usually from the ground. The fruit carries well, but will not keep for any length of time, as it shrivels up. It is principally used as a fresh fruit, though it is also made into jam or jelly, and it often forms part of a fruit salad, taking the place of the granadilla. It has few pests, and is one of the easiest fruits to grow.

Passion Fruit, Redland Bay — Showing method of culture (1) and part of a vine in fruit (2).

CUSTARD APPLES.

Under this heading I will include all the Anonas, such as the sour sop, sweet sop, bullock's heart, and cherimoya. The sour sop is purely tropical, and is very sensitive to frost, but the other species are by no means so tender, and can be grown anywhere along the coast where the soil is suitable, as well as at many inland places. All the species produce very fine fruits, that vary somewhat in shape, in the roughness of the skin, and in size. The sour sop is the largest, and attains a size of 6 to 8 lb. The fruit is covered with soft spines, and is of an irregular oval, or even pyriform, shape. It ripens very soon after it is gathered, consequently cannot be sent any distance. It is a pleasant fruit of an aromatic sub-acid flavour. The pulp surrounding the seeds is of a woolly consistency, and this is surrounded by a custard-like mass which is much appreciated by those who have acquired a liking for it. It is a comparatively uncommon fruit, and is confined to the tropics.

The sweet sop is the commonest of the Anonas, and is grown throughout a considerable part of coastal Queensland. It is usually of an irregular roundish shape, very full of seeds, which are surrounded by a custard-like pulp of very pleasant flavour. It is usually a heavy bearer, and is the variety most commonly met with in our fruit stores. The tree is hardy and is easily grown.

The bullock's heart is a stronger-growing variety than the previous one, the fruit is larger, and, as its name implies, heart-shaped. It is also fairly seedy, the pulp of a light-brown colour, and more gritty, and not, in my opinion, of first-rate quality. It is most commonly grown in the North, where it is a very hardy and prolific tree.

The cherimoya is the best of the custard apples. The tree is a strong grower, with large handsome leaves, but, as a rule, it is not a very heavy bearer. There are many varieties, the fruit of which varies considerably in size and shape, and the skin is sometimes smooth and sometimes warted, or even covered with short soft spines. It has usually comparatively few seeds, and these are surrounded by a rich custard-like pulp, which in the better kinds is of very fine flavour, and is generally much liked. The fruit is not a good keeper, still, given careful handling and packing, it can be kept for nearly a week. All custard apples are easily raised from seed, but the better varieties are propagated by grafting strong seedlings with wood taken from a tree producing fruit of especial merit. Any good fruit soil will grow them, and they do not require any especial treatment.

Custard Apples, Brisbane District.

There are still a large number of tropical fruits that I have not mentioned, but space will not permit of my giving them more than a passing notice, as they are not of any great value from a commercial standpoint at present. Of these fruits the litchi, whampee, averoha, longan, vi-apple, and Chinese mangosteen are practically confined to the North. The guava, of which there are many species, grows anywhere; in fact, it is a pest in many cases, taking complete possession of the land. It is not cultivated to any great extent, as it grows so readily without, and, further, it harbours several pests whose presence it is desirable to remove from the orchard. It is a useful fruit for home consumption, as it stews well, makes an excellent jam, and its jelly is one of the best.

The rosella, a species of hibiscus, is an annual fruit that is grown to a considerable extent in several parts of the State, and is used for pies, jams, and jellies. The latter is remarkably good, equal to that made from the red currant of colder climes, and will no doubt become an article of export at no very distant date. The fruit also dries well, and makes an excellent pickle. It is raised from seed, the young seedlings being set out in well-prepared land when all danger of frost is past. It is a rapid grower, and forms a bush some 4 feet across by 4 or 5 feet high. It is a heavy bearer, and the fruit meets with a ready sale. To do well, the plants require a warm, free, well-drained soil, as they do not thrive where there is any stagnant water at or near the roots.

The avocado or alligator pear is not grown to any extent, though it thrives well, particularly to the north of the tropic of Capricorn, and can also be grown successfully as far south as the New South Wales border. It is a fruit that deserves to be cultivated to a much greater extent than it is at present, and once it becomes better known I have no doubt that it will be planted in considerable numbers, and prove a very welcome addition to our already long list of fruits, as it is unequalled, in my opinion, as a salad. As far as my experience goes, it is likely to become a profitable fruit to grow, as once persons acquire a liking for it, they become very partial to it, and eat it whenever they can get it.

In addition to purely tropical fruits a number of semi-tropical fruits are grown on our eastern seaboard, but are not entirely confined thereto, as many of them are cultivated to a considerable extent in some parts of our coastal and inland tablelands, particularly in sheltered positions. Under the heading of semi-tropical fruits, all kinds of citrus fruits, persimmons, loquats, date palm, wine palm, pecan nut, Brazilian cherry, Natal plum, ki-apple, and many other fruits are included, as well as several fruits that more properly belong to the temperate regions, such as Japanese plums, Chickasaw plum, peaches of Chinese origin, figs, mulberries of sorts, strawberries, cape gooseberries, &c. Of all of these the citrus fruits, which include the orange, mandarin, Seville, lemon, lime, grape fruit, kumquat, citron, and pomelo are by far the most important, and are grown successfully over a very large portion of the State, so that we will consider them first.

Sour Sop, Mossman District.

CITRUS FRUIT.

Quite a number of fruits are included under this heading, and all reach a very high state of perfection in this State. The whole of the family, the lemon-shaped citron excepted, is noted for the beauty and symmetry of growth that its trees make, and I know of few more beautiful sights in the vegetable world than a well-kept citrus grove in full bearing. Take the common round orange as an example, its well-balanced and evenly grown head, its dark glossy green foliage, its wealth of white blossoms, which perfume the whole neighbourhood, or its mass of golden fruit between its dark-green leaves, render it one of the most beautiful of fruit trees at all times, but especially so when covered with blossoms or ripe fruit. A typical Queensland grove is even more beautiful than those of many other places, as the vigour and size of our trees, their exceptionally healthy appearance, their dark foliage, and the heavy crop of high-class fruit that they bear, are at once evident to a stranger who has never seen the orange grown under such favourable conditions as are experienced here. The yield is often so heavy that the trees actually bend to the ground with the weight of their fruit, and a stack of props has to be used to prevent the tree from splitting into pieces. Those who have seen the enormous crops of apples that are produced on some trees in Tasmania or the old cider orchards of Devon or Somerset can form an idea of the crops; but the writer, who has seen both, as well as our Queensland trees, has no hesitation in saying that a Queensland mandarin can give points to either as a heavy cropper; in fact, if it has a fault, it is its proneness to overbear, particularly when young. This all tends to prove how well adapted Queensland is to the growth of citrus fruits, and were I asked to select a country particularly suited to their culture I should have no hesitation in naming this State, as I know of nowhere where their culture can be carried out with less trouble, or where the trees will produce better fruit or heavier crops. Queensland may well be termed the home of citrus fruits, as we have no less than three native species which are indigenous to the State, and are by no means uncommon in our scrubs. Their presence gives unmistakable proof of the suitability of this State for the culture of fruits of the same family, so that I think a short description of these native species may not be out of place, but will be of some interest to my readers.

Young Orange Orchard (6 years old) on scrub land, near Mapleton, Blackall Range. Showing the standing scrub in the background.

Citrus australis, the native orange or lime, is both the largest and most common. It grows into a large tree, having a diameter of 15 to 18 inches in the trunk, and a height of 60 feet or more. It produces a quantity of thick-skinned acid fruit, of from 2 to 3 inches in diameter. The skin is full of a resinous sap, and the fruit is of little value. It is a slow-growing tree, though, as just mentioned, it attains a considerable size, is very hardy, and produces a quantity of fruit. Its slow growth, when young, has prevented its use as a stock on which to work improved varieties, but I have no doubt it would make a very hardy stock that would be distinctly disease-resistant.

The second variety is *Citrus australasica*, the so-called finger lime, a thorny bush, producing a fruit of from ¾ to 1 inch in diameter, and 3 to 4 inches long. The fruit has a thin skin, and contains an agreeable acid pulp that varies in colour, in some specimens being of a reddish tinge that resembles the pulp of a blood orange. These two varieties are met with in the Southern part of the State, but the third is a Northern species, to which Mr. F. M. Bailey, our Colonial Botanist, has given the name of *Citrus inodora*, the North Queensland lime. It is met with in the scrubs of the Russell River, and is described by Mr. Bailey as bearing a greater resemblance to the cultivated species than the two former varieties. It produces a fruit over 2 inches long by 1¼ inches in diameter, having a thin rind and a juicy pulp of a sharply acid flavour, so that even in its wild state it is a desirable fruit, and takes the place of the cultivated lemon. Where native species flourish as they do here, there is every probability of cultivated species thriving equally well, and this is found to be the case in practice.

A young Orange Orchard, Woombye District.

No fruits are more generally distributed or have a wider range in this State than those of the Citrus family, as, with the exception of the colder parts of the Downs, where the winter temperature is too low, the Gulf country, and the dry Western districts, where there is no water available for irrigation, they can be grown from one end of the State to the other, provided that they are planted in suitable soil, and that, in the drier parts, there is an available supply of suitable water with which to irrigate them during the prevalence of long dry spells. The country adjoining the eastern seaboard, extending from the Tweed River in the South to Cooktown in the North—a distance of about 1,100 miles, and extending inland for nearly 100 miles—is naturally suited to the growth of citrus fruits, and there is probably no country in the world that is better adapted to, or that can produce the various kinds of these fruits to greater perfection or with less trouble, than this portion of Queensland. Of course, the whole of this large area is not adapted for citrus culture, as it contains many different kinds of soils, several of which are not suitable for the growth of these fruits, and there is also a large extent of country which is too broken and otherwise unsuitable. At the same time there are hundreds of thousands of acres of land in this area in which the soil and natural conditions are eminently suited to the growth of citrus fruit, and in which the tenderest varieties of these fruits may be grown to perfection without the slightest chance of their being injured by frost; and where the natural rainfall is such that, provided the trees receive ordinary care and cultivation, there is seldom any necessity for artificial irrigation. At the present time there are hundreds of citrus trees growing practically wild in different parts of the coastal country that are in vigorous health and producing heavy crops of good fruit, even though they are uncultivated, unpruned, unmanured, and have to hold their own against a vigorous growth of native and introduced shrubs, trees, and weeds. When the orange, lime, citron, or common lemon become established under conditions that are favourable for their proper development, they apparently become as hardy as the indigenous plants, and are able to hold their own against them, thus showing how well the climate and suitable soils of coastal Queensland are adapted for the cultivation of citrus fruits. The commercial cultivation of citrus fruits is at present practically confined to this coastal area, the most important centres, starting from the South, being Nerang, Coomera, Redland Bay, Brisbane, Enoggera, Gatton, Grantham, Toowoomba, North Coast line from North Pine to Gympie including the Blackall Range and Buderim Mountain; the Wide Bay district, including Maryborough, Tiaro, Mount Bauple, Gayndah, Pialba, and Burrum; the Burnett district, including Bundaberg and Mullet Creek; the Fitzroy district, including Rockhampton and Yeppoon; Bowen, Cardwell, Murray River, Tully River, Cairns and district, Port Douglas, and Cooktown. In addition to these districts a few citrus fruits are grown at Mackay, Townsville, and several other places. Citrus fruits are also grown further inland, but their cultivation here is largely dependent on the ability to supply the trees with suitable water for irrigation during dry spells. Frosts have also to be taken into consideration, for, though the days are warm, the temperature often falls considerably during the night, owing to the great radiation, and citrus-trees in districts like Roma, Emerald, &c., are liable to injury thereby. West of Emerald, at Bogantungan, Barcaldine, and other places, citrus fruits do very well with irrigation. Some of the finest lemons, Washington Navel, and other improved varieties of oranges are grown here to perfection, the lemons especially being of high quality, and curing down equal to the imported Italian or Californian article. The soil in many of the inland districts is well suited to the culture of citrus fruits, and when the trees are given the necessary water, and are uninjured by frost, they produce excellent fruit. I stated, some short distance back, that there is probably no country in the world that is better adapted to the cultivation of or that can produce the various kinds of citrus fruits to greater perfection or with less trouble than the eastern seaboard of Queensland. To many of my readers this may seem to be a very broad statement; but I am certain that, if suitable trees are planted in the right soil and under favourable conditions, and are given anything like the same care and attention that is devoted to the culture of citrus fruits in the great producing centres for these fruits in other parts of the world, we have nothing to fear either as regards the cost of production or the quality of the fruit produced. In order to exemplify this, it may be interesting to compare our capabilities with those of the principal citrus-producing districts north of the equator. To begin with, I will take Florida, which more nearly approaches our climatic conditions than any other citrus-growing country that I know of, and which is noted for the excellence of its citrus fruit, and we find that we have all its advantages except that of proximity to the world's markets, without its disadvantages. We have a better and richer soil, requiring far less expensive artificial fertilisers to maintain its fertility, and at a very much lower price. We can grow equally as good fruit; in fact, it is questionable if Florida ever produced a citrus fruit equal in quality to the Beauty of Glen Retreat Mandarin, a Queensland production. We get as heavy, if not heavier, crops, and our trees come into bearing very early. We have no freeze-outs similar to those which have crippled the industry in Florida so severely in the past that many of their wealthy growers are actually covering in whole orchards of many acres in extent as a protection from frost. This covering-in is accomplished by means of a framework of timber having slat-work or panel sides and tops — in fact, by enclosing their orchards in a huge elaborate bush-house, which is further protected by the heat produced by six large heating stoves or salamanders to each acre of trees enclosed. If it pays the Florida growers to go to all this expense in order to prevent freeze-outs and to produce first-class fruit, surely we can compete with them when a seed stuck in the right soil under favourable conditions will produce a strong, vigorous, healthy tree, bearing good crops without any attention whatever.

An Orange Orchard, near Woombye.

Orange Trophy in the Moreton District Exhibit at the Brisbane Exhibition.

In comparing Queensland with the citrus-producing districts of Southern Europe, we have the advantage of better and cheaper land, absence of frost, more vigorous growth, earlier maturity of the trees, and superior fruit; but with the advantage of cheaper and more skilful labour, especially in the handling and marketing of fruit, and proximity to the world's markets in their favour.

As compared with California, our soil is no better than theirs, but it costs much less, and their citrus industry is dependent on artificial irrigation, their natural rainfall being altogether inadequate for the growth of citrus fruits. Californian conditions more nearly approach those of our inland districts, such as Barcaldine, with the exception that the only rainfall in California is during the winter, whereas in Barcaldine and similar districts the heaviest fall is during the summer months, but, in both, the successful culture of these fruits depends on irrigation.

In Jaffa, also, where the oranges are of large size and extra quality, the trees have to be carefully irrigated and manured, as these operations are found to be essential to the production of marketable fruit.

These few instances show how favourably the conditions prevailing in Queensland compare with those of the great citrus-growing districts of Europe and America, especially in the matter of soil and climate, and I feel confident that, if the industry were taken up in the same business-like manner that it has been done in California and Florida, we could easily hold our own against any part of the world. In comparing Queensland with the rest of the world we have the advantage—also shared by New South Wales and South Africa—of ripening our fruit at a time of the year which is the off season in the citrus-producing countries to the north of the equator, so that our fruit does not clash with theirs, their ripening period and ours being at different times of the year. As regards our Australian market, our fruit ripening earlier than that of the Southern States, we are enabled to dispose of a considerable portion of our crop in the Southern markets before the local fruit is ready for gathering. This gives us three markets—first, a local one; secondly, a Southern one; and, finally, when this demand is supplied, an oversea market to Europe, America, and the East.

When grown under favourable conditions, citrus-trees are heavy bearers in this State, it being no uncommon thing to meet with seedling or worked orange-trees of from ten to twelve years of age producing over twenty cases of marketable fruit to the tree, averaging about 10 dozen medium-sized fruit.

Bunch of Valencia late Oranges, Blackall Range District.

Washington Navel Orange Barcaldine District, Central L

Citrus-trees of all kinds, particularly worked trees, come into bearing very early, and the returns obtained from an orchard rapidly increase. The illustrations give a good idea of the rapid growth, and a fair one of the crop of fruit the young trees are bearing, but the following examples, taken at random for the crop that was marketed in January, 1906, will show better how our trees bear:—

Mr. A., Blackall Range, marketed 7¼ cases per tree from a row of twenty-five Beauty of Glen Retreat Mandarins, planted April, 1900. A return of £1 10s. per tree.

Mr. B., from the same district, averaged 7 cases of Washington Navel Oranges per tree from trees six years old, which realised £1 15s. per tree, and 8 cases of Beauty of Glen Retreat Mandarins from trees of the same age. The navels were large, and averaged 5 dozen per case, and the mandarins 10 dozen per case.

Mr. C, another district, averaged 6 cases of Valencia Late Oranges, from trees six years planted, and 10 cases per tree from Emperor Mandarins, nine years old.

One twelve years old orange-tree in this district produced over 25 cases of fruit.

Mr. D., same district as last; Washington Navels averaged 10 cases per tree, ten years planted, and have borne regular crops since three years old.

Numerous other cases might be given, but the above are sufficient to show the earliness at which our trees bear, and the crops they yield. Trees in full bearing often yield up to 40 cases, but these are usually old seedlings, which bear a very heavy crop one year and a comparatively light crop the next. All the instances I have quoted are from worked trees, which are found to give the most regular and constant yields. Until quite recently, citrus-trees were almost entirely grown from seed in this State, with the result that we have a very large number of types, and many crosses between different species. This was not advisable, as a uniformity in type is desirable for marketing, hence the greater number of trees now being planted are of selected varieties of proved merit. Many of the seedlings have produced most excellent fruit, but a seedling has usually the disadvantage of being very full of seeds, and having a lot of rag (the indigestible fibre round the pulp) as compared with the worked varieties, which have either no seeds or very few seeds and little rag. Seedlings are also of many types, and they produce a lot of small fruit, thereby making an uneven sample, whereas worked trees produce fruit even in size and quality. Seedlings are probably the hardiest, and will stand the most neglect, but experience is showing that worked trees are the most profitable to grow. The growth of all kinds of citrus-trees from seed is a very simple matter, all that is necessary being a well-prepared seed bed of friable soil that is partially shaded from the heat of the sun, so as to protect the young plants. Selected, fully ripe fruit from well-grown, prolific, healthy trees is taken, and the seeds sown in rows in the seed bed, or broadcast when weeds are not likely to be any trouble. Fresh seed germinates quickly, and the young plants are soon ready to be transplanted into the nursery bed, where they are either worked over or allowed to remain seedlings. At twelve months old, from seed, a tree will have a stem-diameter of about ¾-inch, and a height of 3 to 4 feet, a growth about twice that made in the Southern States.

The general remarks I have given respecting our fruit soils apply with equal force to those best adapted for citrus culture—viz., they must possess perfect drainage, and be of a friable nature. We are growing most of the best varieties of citrus fruit, the original trees from which they are now being propagated having been introduced into the State from the most celebrated citrus-producing districts in the world, and, as stated and shown by the accompanying illustrations, they are all doing well.

The Washington Navel, the variety of orange most commonly grown in California, does remarkably well on our rich volcanic scrub soils, where it has proved itself a regular bearer of high-class fruit. The Mediterranean Sweet Orange, Valencia Late, and Jaffa also do well in many parts, the Valencia Late adapting itself to most districts. Many other kinds of oranges are grown, but the varieties mentioned are some of the best, and are the ones now being planted in the greatest quantity.

Spray of Orange Blossom.

In mandarins, all kinds do remarkably well, and I never saw this fruit produced to greater perfection in any part of the world than it is in Queensland. The varieties most commonly grown are: The Emperor or Canton, the Scarlet or Scarlet Emperor, Thorny or Tangerine, and Beauty of Glen Retreat, though there are many types of seedlings in addition to these well-known sorts. The grape fruit which is now so popular in America does well, but, so far, has not taken on in our markets. Citrons grow practically wild, and produce good fruit, for which there is a limited demand for peel. Their cultivation could be extended with ease were there a better demand for peel. The Seville Orange, which is used for the manufacture of marmalade, is an exceptionally hardy and prolific tree, and, were it required, we could easily grow enough of this fruit to supply the world. Lemons do best inland, or at an elevation of some 2,000 feet above sea-level, as this fruit is apt to become too coarse in the skin when grown in a humid climate. In suitable localities very good fruit can be grown, which compares very favourably with the European or American grown fruit.

The lime does well in the more humid districts, taking the place of the lemon, and one variety — the Tahiti — has proved itself to be a heavy and regular bearer. The West Indian lime, from which the lime juice of commerce is made, is very easily grown, particularly in the more tropical parts, where it is often met with growing in an entirely uncultivated condition, and bearing heavy crops of fruit. Kumquats are easily grown, and are heavy bearers, and all the different types of pomelos or shaddocks do well. Seedlings of the latter are very hardy, as they are deep-rooted plants that stand dry weather well and are, consequently, not liable to injury during dry spells. There is very little demand for the fruit, but I am of opinion that the seedlings will prove to be of value as stocks on which to work our best kinds of oranges.

The culture of all kinds of citrus fruits, when grown in suitable soil, is by no means difficult, as it consists mainly of keeping the land well stirred and keeping down all weed growth during dry spells, the keeping of the trees well pruned out in the centre, and the keeping in check of all diseases, both insect and fungus. Although citrus fruits are subject to many pests, they are for the most part easily kept in check by either spraying or cyaniding, or both, provided that reasonable care is taken, and the pests are destroyed before they have obtained control. Taken as a whole, our citrus fruits are remarkably clean, and compare more than favourably with those grown in the Southern States. The culture of these fruits is extending rapidly, with a corresponding increase in production, but, despite this, our prices have been better during the past season than for some years, as the quality of our fruit is such that it will command a good market. When properly handled, it has good keeping qualities, and I have no doubt that we will, in time, be able to supply the markets of the Old and New Worlds with good fruit, in the best of condition, at the time of the year that their markets are bare of locally-grown citrus fruit.

There is a good opening for the growth of citrus fruits in this State, as the writer knows of no country where they do better, where they can be produced with as little trouble and expense, where they can be successfully grown over such a large area, or where the soil and climate is more suited to the production of fruits of the highest quality as in Queensland.

Lisbon Lemon, Esk District.

THE PERSIMMON.

This exceedingly handsome fruit of Japanese origin is grown to a high state of perfection in this State, particularly in the coastal districts south of the tropic of Capricorn. It is a fruit of comparatively recent introduction, the oldest trees being less than thirty years of age, but has already become widely distributed, as well as a favourite fruit amongst many. It is a very showy fruit when well grown, but must be thoroughly ripe before it is eaten, as, if not, it is extremely astringent, and anyone who has tackled an unripe fruit has no wish to repeat the experience in a hurry. There are many varieties of this fruit, some of which are seedless, and others more or less seedy. The seedless kinds are usually preferred, as, as well as being seedless, they are the largest and handsomest fruit. The different kinds vary considerably in the size of tree, habit of growth, foliage, size and colour of fruit, &c. All are easily grown, and most kinds are good and regular bearers. They do well on any fruit soil, and some of the dwarf-growing kinds are well adapted for growing in private gardens, on account of the small amount of room they take up. The trees are deciduous, and, as a rule, are not much troubled with pests. So far, the use of the fruit is confined to its consumption fresh, though in Japan it is dried in a similar manner to apricots or peaches.

Persimmons.

THE LOQUAT.

A handsome evergreen tree, that can be grown in the more Southerly coast districts, in the foothills of the coast range, and on the coast tablelands. There are several types of the fruit, whose chief value consists in that it ripens its fruit in early spring, when there is a shortage of stone fruits, and that it withstands wind well, so makes a good break for the protection of exposed orchards. Its cultivation is not extensive, nor is it likely to become so.

Fruit of Loquat (½ natural size).

THE DATE PALM.

Although this extremely valuable fruit is grown in this State more as an ornament than for its commercial value, there is nothing to prevent its culture on a scale sufficiently large to supply the Australian requirements. It is grown in many places along the coast, as well as in the foothills country of the coastal range, but it does best in situations that more nearly resemble its natural habitat—viz., in districts having a hot dry air, a deep sandy loam or sandy soil, and a good supply of moisture in the soil. This latter condition does not occur naturally, but can be supplied artificially in our Western lands, where there is a good supply of artesian water of a quality suitable to the plants' requirements. Here the date palm thrives, and produces huge bunches of fruit. Little, if any, cultivation is necessary when once the palm is firmly established; provided it has an ample but not excessive supply of moisture, it is able to take care of itself.

The date palm is a diœcious plant—that is to say, the male organs, or stamens, are produced on one plant, and the female organs, or pistils, on another, and this necessitates the growing of the two sexes in proximity to each other, in order that the female flowers may be fertilised and produce perfect fruit. This is best accomplished artificially, the pollen from a fully developed bunch of male flowers being shaken over the bunch of female flowers. Infertile fruit contains no seeds, and is of small size and inferior quality, whereas the fertile fruit is both large and good.

The date palm is a handsome ornamental plant, and in the hot and dry Western districts, where it thrives best, it forms a splendid shelter from the sun for both man and beast. So far, very little attention has been given to its growth, few persons knowing how to fertilise the flowers or even taking the trouble to see that they have plants of both sexes. There is no reason why this should be so, as there would be a good local demand for the properly-cured fruit, and I believe that, were its culture carried out in a thorough business manner, it would become a profitable industry, and one capable of supplying our Australian market.

Date Palms in fruit at Barcaldine.

THE PECAN NUT.

Another little-known fruit which does well in this State. It belongs to the hickory family, and closely resembles the walnut. There are trees now growing in the Maryborough district that are some 15 inches in diameter at the trunk, and from 40 to 50 feet in height, that bear regular and heavy crops of nuts, and that have stood drought and been under flood. For years the trees have received no cultivation, and they have shown themselves to be as hardy as the adjacent indigenous trees. The trees are easily raised from seed, and come into bearing in about eight years. Like all nut fruits, it is advisable to set the nut where the tree is to remain permanently, if it is possible to do so, as it produces a very deep taproot, with few laterals, and is consequently difficult to shift. The soil on which it does best is an alluvial loam, and, if possible, it should not be more than 30 feet to water, as the tree, being a very deep rooter, will penetrate a free soil to that depth. It will do on other free loamy soils, but will not make the same growth as when planted in free alluvials. It has been tested in several parts of the State, and it is probable that it will be found to thrive over a considerable area of the coastal and coastal tablelands districts. It produces an olive or acorn shaped nut, having a thin shell, and of a flavour closely resembling that of a good walnut, and will be a valuable addition to our list of nut fruits once it becomes better known.

Date Fruit (natural size).

JAPANESE PLUMS.

All varieties of this fruit thrive well and bear heavily in the more Southerly part of our coast country, as well as on the country immediately adjacent to it, the coastal tablelands, and several other parts of the State. The trees are rapid growers, come into bearing very early, and often bear enormous crops of fruit. They are good fruits for home consumption or for the fresh-fruit trade, but are not equal to European varieties of plums for preserving, drying, or jam-making. In this State they have one very great drawback, and that is their liability to the attack of the fruit fly, a pest that very frequently destroys the entire crop. For home use they are, however, a very useful fruit to grow, provided that the trees are kept dwarf, so that they can be covered with a cheap mosquito netting as a protection from the fly, as they are very easily grown, are by no means particular as to the kind of soil on which planted, and are heavy bearers.

CHICKASAW PLUMS.

This family of American plums does well in the same districts as the Japanese varieties just dealt with, but has the advantage of being resistant to the fruit fly. The trees are usually more or less straggling growers, the fruit is of small size, but good for cooking or jam-making. One or more of the varieties of this plum are bad setters, though they blossom profusely, but this may be overcome either by working two varieties which bloom at the same time on to the same stock, or by planting varieties that bloom at the same time together, as the pollen from the one will set the fruit of the other. It is a good plum for home use or marketing, despite its small size, as it is easily grown, requires little attention, and is not over particular as to soil.

CHINESE PEACHES.

Peaches of Chinese origin thrive well on the coast, and are extremely hardy. The fruit is not, as a rule, of high quality when compared with that of the Persian varieties, but their earliness and ease with which they can be grown causes them to be planted by many who have small gardens. Like the Japanese plums they are, however, very subject to the attack of fruit fly, and require to be kept dwarf and covered in a similar manner if any good is to be got from them. On the coast, they are practically evergreen, as they never lose their leaves entirely, and are in blossom during the winter. When grown on the tablelands, this early blossoming is a disadvantage, as the blossoms are liable to be injured by frost, but in these districts peaches of Persian origin can be grown instead.

FIGS.

Several kinds of figs can be grown successfully in the Southern coast districts, the first crop ripening before Christmas, but the second or main crop is often a failure, owing to the fact that it ripens during our wet season, and the fruit consequently sours and bursts. As one recedes from the coast, the fruit does better, and is less liable to injury from excessive wet. The coastal tablelands and the more Western Downs grow it well, and the trees, when planted on soil of a rich friable nature, grow to a large size and bear heavily. Many varieties are grown, which are used fresh or converted into jam, but no attempt has been made to dry them, though it is possible that this industry may eventually be found profitable in the drier parts of the State, where there is water available for the trees' use at certain periods of the year, but not during the fruiting period, as it cannot well be too dry then if a good quality of dried figs is to be turned out. This fruit is easily grown, and is not at all subject to serious pests, so that anyone who will take reasonable care can produce all that is required for home use or local sale, as its softness renders it a difficult fruit to ship long distances in a hot climate.

THE MULBERRY.

This is one of the hardiest fruits we have, one of the most rapid growers, and one of the most prolific. There are several varieties in cultivation, and those of Japanese or Chinese origin will grow from the coast to the interior, and thrive either in an extremely dry or humid climate. The common English or black mulberry does not do too well as a rule, though there are many fine trees scattered throughout the State, but the other sorts are as hardy as native trees. The fruit is not of any great value, still, as it is so easily grown, it finds a place in most gardens, and in time of drought the leaves and young branches are readily eaten by all kinds of stock, so that it is a good standby for stock as well as a fruit.

THE STRAWBERRY.

To those who have been accustomed to look upon the strawberry as a fruit of the purely temperate regions, it will be somewhat of a revelation to know that exceptionally fine fruit can be grown right on the Queensland coast, and well within the tropics, and that on the coast, between the 26th to the 28th degrees of south latitude, we are probably producing as fine fruit and obtaining as heavy crops as are produced in any of the older strawberry-growing countries. Not only this, but that we are able to supply the Southern markets of Australia with finer fruit than they can produce locally, and at a time of the year that they cannot grow it. As I have already mentioned when dealing with other fruits, one thing that particularly impresses strangers is the early age at which our fruits come into bearing. This is borne out in the case of the strawberry to a marked degree, as runners set in April fruit in July, and often earlier, and will continue to bear, given reasonable weather, right up to Christmas or even longer. New plants are set out every year, and the plantation is seldom allowed to stand more than two years, as the young plants produce the finest fruit. There is a good demand for the fruit, the larger berries being packed in flat cases holding a single layer of fruit, as shown in the illustration, and being sold for consumption fresh, whereas the smaller berries are packed in kegs and sent direct to the factories for conversion into jam. The strawberry grows well on various soils, but does best with us on a rich loam of medium texture, of a reddish-brown or even black colour. It should be planted in districts that are free from frosts where early fruit is desired, as frosts injure the blossoms, but where jam fruit only is wanted this is not so necessary. The land requires to be thoroughly well prepared, and the plants are usually set out in rows about 2 feet apart, with the plants about 1 foot apart in the row. Under favourable conditions they grow very rapidly, and soon start flowering. Their cultivation is usually confined to comparatively small areas of 2 or 3 acres in extent, as the labour of picking and packing is usually done by the grower himself with the assistance of his family. They are often planted between the rows of trees in young orchards, thus bringing in a return whilst the trees are coming into bearing, and helping to keep the pot boiling. They grow well on our coastal scrub lands, and have proved a great assistance to many a beginner, as one has not long to wait before obtaining a return.

Strawberry Garden, Mooloolah District.

The productiveness of this fruit in Queensland is phenomenal, as high as 5 tons of berries having been taken off 1 acre in a single season. There are many varieties of strawberries in cultivation, some of which have been produced locally from seed, and have turned out extremely well, being of better flavour, stronger growers, and heavier bearers than introduced varieties—in fact, local seedlings have adapted themselves to local conditions, and stand our climate better than those varieties which are natives of colder countries.

Marguerite Strawberry.

Marguerite Strawberry packed for market.

The case berries, which are used for fresh consumption, fetch a fair price, especially early in the season, but jam fruit sells at an average of 2-1/2d. per lb., at which price it pays fair wages, but is not a bonanza. As a rule the plants are very healthy, and any fungus pests to which they are subject, such as leaf blight, are easily kept in check by spraying, a knapsack pump being used for this purpose.

The ground is kept well worked and free from weeds, whilst the plants are fruiting, and occasionally the ground is mulched, as is the case in the plot shown in the illustration. No special knowledge is necessary for their culture, but, at the same time, thorough cultivation and careful attention to details in the growing of the plants make a considerable difference in the total returns.

Forman's Strawberry, Brisbane District.

CAPE GOOSEBERRY.

This Peruvian fruit, introduced into this State *viâ* the Cape of Good Hope, hence its name, has now spread throughout the greater part of the tropical and semi-tropical portions of Queensland. Its spread has largely been brought about by the agency of fruit-eating birds, that have distributed the seeds widely by means of their castings. It is one of the first plants to make its appearance in newly burnt-off scrub land, and often comes up in such numbers as to give a full crop of fruit. In other cases it is usual to scatter a quantity of seed on such land, so as to be sure of securing a plant. No cultivation is given; the plant grows into a straggling bush bearing a quantity of fruit which is enclosed in a parchment-like husk. The fruit is gathered, husked, and is then ready for market. The bulk of the fruit is grown in this manner, and as it can be grown on land that is not yet ready for any other crop (grass or maize excepted) it is a great help to the beginner, as a good crop and fair prices can usually be obtained. The name "gooseberry" is somewhat misleading, as it is not a gooseberry at all, is not like it, nor does it belong to the same natural order. It is a plant belonging to the order Solanaceæ, which includes such well-known plants as the potato, tomato, tobacco, &c., and altogether unlike the common gooseberry, which, by the way, is one of the fruits that we cannot do much with. In addition to being grown in the wild manner I have described, it is occasionally cultivated in a systematic manner, somewhat like the tomato, but not to any extent; growers preferring to depend on it as a first return from newly fallen and burnt-off scrub land. As a fruit it meets with a very ready sale, as it is one of the best cooking fruits grown; plainly stewed and served with cream, made into puddings or pies, or converted into jam, it is hard to beat. The jam has a distinct flavour of its own, one that one soon becomes very partial to, besides which it is an attractive-looking jam that, were it better known in the world's markets, would, I feel sure, meet with a ready sale at satisfactory rates. The plant is somewhat susceptible to cold, hence it does best in a district free from frost, but it is not killed out by light frosts, only killed back, and its crop put back. Like all plants belonging to the same natural order, it likes a good soil, rich in available potash, and this is probably the reason why it does so well on newly burnt-off scrub, the ashes of which provide an ample supply of available potash.

THE OLIVE.

A much-neglected fruit in this State, as it is also in most English-speaking countries. Few English people are fond of either the fruit or the oil, and yet it is probable that there is no tree that for the space it occupies will produce a greater annual return of food than the olive. A number of trees are scattered throughout the State, some of which are now of large size and fair age, but, so far, practically nothing beyond making a few gallons of oil and pickling a few gallons of fruit has been attempted, and this only in a purely experimental manner.

The present condition of the olive industry is destined to have a wakening up ere long, as a country that can produce this fruit in such quantities and of such a quality as the lighter soils of the Darling Downs is destined some day to be one of the largest producers of olives on earth. Some years since I planted a number of the best varieties of olives—trees obtained direct from California—on the Darling Downs, in land that I considered suitable for their growth, and which was properly prepared prior to planting. The trees here have made a really phenomenal growth, they came into bearing within three years of planting, and have borne steadily ever since. They have proved enormous bearers, and an experimental crushing showed that the oil was of high quality.

There are large areas of similar country to that in which they are planted in different parts of the State, and I feel certain that this really valuable food fruit is bound some day to be a considerable source of our national wealth. So far, the drawback to the growth of olives has been the cost of gathering the fruit and the limited demand for the oil or pickled fruit, but, against this, it has many advantages, one, and by no means the least, of which is its value as a shade and shelter tree on our open treeless plains. It is also a very hardy tree, withstanding drought well, and thriving in land that is too stony for the cultivation of ordinary farm crops. It is a healthy tree, free from most fruit pests other than the olive scale, which can be kept in check by spraying or cyaniding; and last, but not least, it is an ornamental tree whose wood is of considerable value. The olive does best with us in loamy soils of fair depth and basaltic origin, that are moderately rich in lime and potash, and have a fair drainage. A subsoil of decomposed rock answers well. It will, however, do on several other kinds of soil, but it is in the type that I have just described that it does so well, and in which I would recommend its culture on a large scale. It will stand a fair amount of frost as well as great heat, and I have never seen the trees injured by either on our Downs country. I have also seen trees doing well right on the coast, where they have been subject to heavy rainfalls, so that it appears to adapt itself to the conditions prevailing in many parts of our State.

In addition to the fruits I have briefly described, there are several others of minor importance that can be grown successfully, but, as they are not of any great value commercially, I will leave them out, and go on to the fruits of our more temperate districts, as, in addition to growing the tropical and semi-tropical fruits which I have already dealt with, Queensland can also produce temperate climate fruits to a very high degree of perfection.

The fruits of the temperate regions that we are able to grow include the apple, pear, plum, prune, quince, apricot, Persian peach, nectarine, almond, walnut, chestnut, cherry, &c., as well as some of the hardier fruits which I have classed as semi-tropical—viz., the Japanese plum, persimmon, Chickasaw plum, strawberry, &c. The districts adapted for the growth of the distinctly temperate fruits are mostly situated in the Southern portion of the State, and at an elevation of from 2,000 to 3,000 feet above sea-level—districts having a warm summer but a comparatively cold winter, during which frosts are by no means uncommon, but where snow rarely falls; a healthy climate, with warm days and cool nights, to which many visitors go during the heat of summer, when the humidity of the coast is somewhat trying to persons not naturally robust. The Downs country, particularly its southern or Stanthorpe end, is the most suitable; the soil is mainly of granitic origin, and is very suitable for the growth of apples, stone fruit, and grapes, but the latter I will deal with by themselves later on. The country is by no means rich from an agricultural standpoint, and is considerably broken, but, as already stated, it is admirably adapted for the growth of fruit, and within the last ten years at least 100,000 fruit trees, mostly apples, plums, and peaches, have been planted out and are doing well. The Stanthorpe show, which is held annually during the month of February, is always noted for the excellence of its fruit exhibits, which would be hard to beat, both for size, quality, and appearance. The fruits ripen earlier than similar varieties grown in the Southern States, hence supply our markets at a time when there is little outside competition, and, consequently, meet with a ready sale at fair prices. The fruit grown in the largest quantity is the apple, so I will deal with it first.

THE APPLE.

As a description of this well-known and universally used fruit is entirely superfluous, I will confine my remarks to the types of fruit grown, and their method of growth. Owing to the fact that our fruits ripen much earlier than similar varieties in more southern parts of Australia, we have gone in largely for early varieties of apples, both for cooking and table use, but have not confined our attention to them entirely, as good-keeping sorts are found to do equally well, and have been shown at the annual exhibition that is held in Brisbane during August, in perfect condition, showing that the fruit has good keeping qualities. The soil on which the apple is mostly grown is largely composed of granitic matter, and is of a sharp, sandy, loamy nature, often of a gritty character. It is usually rich in potash, the predominating felspar being orthoclase, but somewhat deficient in nitrogen and phosphoric acid. It is usually easy to work, of fair depth, and retains moisture well when kept in a thorough state of tilth. The trees are usually planted at from 20 to 25 feet apart each way, when they are either one year or two years old from the graft or bud. They are headed low, so as to shade the ground from the heat of the sun, and also so as to facilitate the handling of the crop when grown, as well as to prevent their swaying about with the wind. The trees make a rapid growth, come into bearing very early, often bearing a fair crop three years after planting, and fruiting even earlier. The fruit of the early varieties has usually a handsome appearance, but lacks keeping qualities, but the later fruits are both handsome, high-coloured fruit, and good keepers. The trees are not very liable to disease, as, thanks to all varieties being worked on blight-resistant stocks, there is very little American blight (woolly aphis). Scale insects do a certain amount of damage, but are easily kept in check by winter spraying, and codling moth is not bad unless grossly neglected, many orchards being quite free from this great pest of the apple-grower. So far, the growing of apples has been confined entirely to the growing of fruit for the local markets, no attempt having been made to export same. A very small quantity is dried, and a little is used for jelly.

Many varieties of apples have been tested in this State, but growers have found out that it pays them best to confine their attention to comparatively few sorts that have proved to be the best suited to the soil and climate, as a few good kinds are much more profitable to grow than a mere collection of varieties. Many varieties are prone to overbear, and trees of large size have produced enormous crops of fruit, whereas young trees frequently break down under the weight of their crop. The usual plan is to plant a few varieties that ripen in succession, so as to extend the season over as long a period as possible, and not to cause a glutted market at any one time. Early fruits particularly are not noted for their keeping qualities, and a market glutted with such would entail a heavy loss to growers, hence a succession of varieties that suit the district as well as the market is grown.

Nearly all kinds of apples do well, those that are resistant to the attack of woolly aphis are, however, generally chosen in preference, even though they may not be of the highest quality, as their prolificness and freedom from this pest renders them more profitable than varieties of superior quality that are liable to blight, and that are at the same time often somewhat indifferent bearers. It is outside the scope of this paper to go into the question of varieties, but I may mention that such sorts as Irish Peach, Gravenstein, Summer Scarlet Pearmain, Twenty-ounces, Jonathan, Lord Suffield, Rome Beauty, and Prince Bismarck do remarkably well, and many other well-known kinds can be grown to perfection.

Prince of Pippins Apple, Darling Downs District.

THE PEACH.

This king of the temperate fruits grows with us to perfection. The tree is hardy, a rapid grower, comes into bearing early, and is, if anything, inclined to overbear. It can be grown over a considerable part of our coastal and inland downs, as well as the Stanthorpe district, and thrives in many kinds of soil, from light sandy loams of poor quality to rich loams of medium texture or even heavier. In this State, the peach is always grown on peach roots, the desired variety being either budded or grafted on to a seedling peach, and the resulting tree is planted out when it has made one year's growth. No tree is easier to grow, but if the best returns are desired, it requires very careful pruning for the first three years, after which an annual winter pruning is usually all that is necessary. The young tree is such a strong grower that unless it is heavily cut back it becomes top-heavy and breaks to pieces with the weight of fruit, but when hard cut back for the first two years, so that it has a good main stem and strong primary branches, it will form a strong tree, and stand up well under a heavy crop of fruit. The strong growth it makes necessitates heavy pruning when large fruit is desired—and it is large showy fruit which sells best here—as were the tree allowed to go unpruned, it would bear enormous numbers of fruit, many of which would be of small size. Growers now realise this, and many of our orchards are well pruned, whereas a few years since the trees were allowed to grow pretty much as they like.

The peach remains profitable much longer here than it does in California, as the trees do not wear out so quickly, the roots remaining sound up to the last, so that, unless the top is too far gone, the life of the tree may usually be extended for several years by heading hard back and forming an entirely new head to the tree. Trees in full bearing often produce fully 1,000 lb. weight of fruit in a single season. This is, of course, very much above the average, but by no means exceptional. When in their third season, they should bear enough to pay for all working expenses.

A very large number of varieties have been tested in Queensland, most of which do well, but, as in the case of apples, we find from experience that it is best to stick to a few kinds, and those that have proved to be most suitable to our soil and climate, rather than to experiment with a large number of varieties.

The usual plan is to plant a number of varieties that ripen in succession, as with the apple, so as to spread the season over as long a time as possible, and to stick to kinds that bear well, look well, and ship well, for appearance will usually beat quality, and fetch more money.

So far, little has been done in the way of utilising the peach, as the demand for the fresh fruit has been equal to our supply. There is, however, no reason why we should not be able to establish and maintain a fair canning and drying trade, should the production overcome the demand for the fresh fruit, as our peaches are of large size, and will can and dry well—that is to say, varieties adapted to those purposes will do so.

The nectarine, which is simply a smooth-skinned peach, does equally well, many varieties bear heavily, and some produce fruit of exceptional merit. I have seen as fine nectarines grown in the Stanthorpe district as I have met with in any part of Australia or America, fruit of large size and the highest flavour, that compared favourably with the finest hothouse-grown fruit of the Old World.

Peach Avenue, Darling Downs District.

THE PLUM.

As already mentioned, plums of Japanese and American origin (Chickasaw) do well in the more coastal districts. They also bear heavily on our coastal downs and more western country, but some kinds of Japanese plums blossom too early for the Stanthorpe district. European plums, however, do well, and are heavy bearers. All kinds do not bear heavily, the freest bearers being those of the damson family—White Magnum Bonum and Diamond type. Prunes also do well. Plums of European origin do best in the coldest districts, but their cultivation is not confined entirely to these, as some varieties thrive well in warmer and drier parts of the country. So far, there has been a ready sale for all the plums we can produce for fresh consumption, excepting some of the smaller plums of the damson type, which have been converted into jam. It is not a fruit, however, in which there is much money, as it is too easily grown in the Southern States, and can there be converted into jam or canned at a lower rate than we can do here, hence our cultivation will be more or less confined to the growing of large fruits for supplying our local markets rather than to the production of the fruit in quantity.

THE APRICOT.

Most varieties of this fruit do well on our coastal downs country in the South, and to a certain extent further west. The trees are very rapid growers, and bear heavily. The earlier ripening fruit usually escapes damage from fruit fly, but the late fruit often suffers considerably.

The apricot does best in a fairly strong rich soil, when it makes a great growth, and bears heavy crops of large-sized fruit. It also does well on sandier soils, which produce a firmer and better-drying fruit. So far, although a number of trees are planted throughout the State, the cultivation of the fruit is mainly confined to the production of table fruit, drying or canning having been carried out to a small extent only. The apricot grows to a large tree, and lives to a good old age. Like the peach, it is a very vigorous grower when young, requiring severe pruning in consequence, but, when once shaped, the trees require little in the way of pruning other than the removal of superfluous branches and an annual shortening in winter.

THE CHERRY.

Queensland is almost outside the limit of the successful growth of this fruit, but not quite, as we produce the first fruit to ripen in Australia, which realises a high price on account of its earliness. Many varieties have been tested, but, so far, no one variety can be said to be a complete success in our climate, nor do the trees grow to the large size or produce as heavily as they do in the Southern States, where the winters are more clearly defined than they are in Queensland. Another drawback to the growth of this fruit is that the soils of our coldest district are not the best of cherry soils. The cherry likes a deep, moderately rich loam, whereas we are growing it mostly on sandy loams of a granitic origin. What fruit we do grow is good, and pays well on account of its earliness, but I do not consider that this State will ever be able to compete with the South in the growth of the cherry.

Litchi, Mossman District.

THE PEAR.

Many kinds of pears do well, but, unfortunately, this fine fruit is very liable to be attacked by fruit fly. It does well generally in the districts that I have mentioned as suitable for the apple, plum, and apricot. The tree is healthy, grows rapidly and to a large size. It comes into bearing remarkably early as compared with the pear in colder climates, and produces excellent fruit. I have grown as good Bartletts here as could be obtained anywhere, and the trees have proved to be good bearers and doers. This fruit does best on deep soils of a medium to strong loamy nature, and of good quality, though it does well in much freer soils, but does not make as good a growth or bear as heavily. It is usually grown on seedling-pear stocks, but the growing of suitable varieties on quince stocks and keeping the resultant trees dwarfed is to be recommended. This method of growing the pear does well here, and dwarf trees can be easily protected from fly, whereas it is practically impossible to deal with big trees, which the pear becomes when grown on pear roots.

THE ALMOND.

This fruit does well in parts of our coastal tableland country, though its habit of blossoming too early in the season renders it very liable to injury from late frosts. The trees do remarkably well, grow rapidly, and bear heavily when the blossoms are uninjured by frost, hence it is a good tree to grow in selected situations containing suitable soil, as it commands a ready sale, and is very little troubled with pests. A free, sandy, loamy soil is best suited to the growth of the almond, and the situation should be well protected from frost. The trees are usually worked on peach stocks, on which they make a very rapid growth. Several varieties should be grown together, as a better set of fruit will be obtained by doing so, most almonds requiring the pollen of another variety flowering at the same time to render their flowers fertile. The almond grows into a handsome, shapely tree, and, when in blossom, an orchard is a sight not easily forgotten, the wealth of flowers being such that it must be seen to be fully appreciated.

The walnut, chestnut, quince, blackberry, raspberry, and one or two other fruits of the temperate regions are also cultivated to a small extent, but are of no great value so far, though there is no reason why the walnut, which does well with us, should not be cultivated to a much greater extent than it is, as there is always a fair demand for the nuts. Blackberries of different kinds have been introduced, and do well, the common English blackberry almost too well, as unless kept in check it is apt to spread to such an extent as to be a nuisance. In addition to the cultivated fruits I have briefly mentioned as growing in Queensland, we have a number of native fruits growing in our scrubs and elsewhere that are worthy of cultivation with a view to their ultimate improvement. Of such are the Queensland nut, a handsome evergreen tree, bearing heavy crops of a very fine flavoured nut. The nut is about ¾-inch in diameter, but the shell is very hard and thick. It could no doubt be improved by selection and careful breeding. The Davidsonian plum is also another fruit of promise. It is a handsome tree of our tropical North coast, and bears a large plum-shaped fruit of a dark purple colour, with dark reddish purple flesh, which is extremely acid, but which is well worth cultivation. Several species of eugenias also produce edible fruit, and there are two species of wild raspberries common to our scrubs. There are the native citrus fruits I referred to in an earlier part of this paper, as well as several other less well-known fruits that are edible.

Tamarind Tree, Port Douglas District.

GRAPE CULTURE.

No work on fruit-growing in Queensland, however small, would be complete without due reference being made to the vine, the last but by no means the least important of our many fruits. Although the cultivation of this most useful and popular fruit has not reached to anything like the dimensions that vine culture has attained in the Southern States, particularly in the production of wine, there is no reason why it should not do so at no very distant future. We have many advantages not possessed by our Southern neighbours in the culture of the grape, the first and most important of which is that our crop ripens so much earlier than that of the South that we can secure the whole of the early markets without fear of any serious opposition. Until quite recently, grape culture was in a very backward state in Queensland, the grapes grown on the coast being nearly all American varieties, which are by no means the best wine or table sorts. A few grapes of European origin were grown on the Downs and in the Roma district, but their cultivation was practically confined thereto. Now, however, things have altered very much for the better. Many good varieties of European grapes have been proved suitable to the coastal climate of the Southern half of the State, and many inland districts other than Roma and the Downs have also proved that they, too, can and do grow first-class fruit both for table and wine.

Grosse Kölner Vine in Fruit, Roma District (Gros Colman).

Picking Grapes, Roma.

Now the culture of the grape extends over a great part of the State, from the coast to the interior; in the latter, its successful growth depending on the necessary suitable water for irrigation, and on the coast to our knowledge of how to keep fungus pests, such as anthracnose, in check by winter treatment and spring spraying.

In the Brisbane district many kinds of excellent table grapes are now grown, which meet with a ready sale, such as the well-known Black Hamburgh of English vineries, the Sweetwater, Snow's Muscat Hamburgh, Royal Ascot, &c., as well as all the better kinds of American grapes, such as Iona, Gœthe, Wilder, &c. A little wine is made, but more attention is given to table fruit.

In the Maryborough, Gympie, and Bundaberg districts, similar grapes are also grown, and do well, ripening somewhat earlier than they do in Brisbane; and in the Rockhampton district, right on the tropic of Capricorn, some of the best table grapes I have seen in the State are produced. Further north a few grapes are grown, but not in any great quantities, and I consider that the profitable cultivation of good table grapes on the coast extends from our Southern border to a short distance north of the tropic of Capricorn and inland to all districts where there is either a sufficient rainfall or a supply of water from artesian bores, or otherwise, to enable them to be grown. Grapes here, as in other parts of the world, like moderately rich, free, loamy soils of good depth, free sandy loams, and free alluvial loams. In such soils they make a vigorous growth, and are heavy bearers. The granitic soils of the Stanthorpe district, that produce such good peaches, plums, and apples, grow excellent grapes, which ripen late. They are of large size, and conspicuous for their fine colour. The sandy soils of Roma and the Maranoa country generally grow excellent wine and table grapes, the latter being of large size, full flavour, and handsome appearance. Wine grapes also do well here, and some excellent wine has been made, both dark and light, natural and fortified. I have no doubt that eventually good rich port and the best of sherries will be produced in this district, as the soil and climate are admirably adapted to the production of these classes of wine. Our difficulty, so far, has been to find out the exact kinds of grapes to grow for this purpose, but now I am glad to say that we are on the right track, and the excellence of Queensland ports and sherries will be a recognised thing before many years are past. There is a big and good opening for up-to-date viticulturists in this State. We have any amount of suitable land at low rates, and, thanks to the generous sun heat of our interior, we can grow grapes capable of producing wines equal to the best that can be turned out by Spain, Portugal, or Madeira. In those districts that do not possess such an extreme climate, such as the coastal downs and the Stanthorpe districts, good wines of a lighter character can be produced, and, as already stated, good wines are now being made on the coast.

A Grape Vine in Fruit, Stanthorpe District.

Madresfield Court Grape.

It is only now that we are beginning to realise the value of the grape to Queensland, as, until our production increased to such an extent that our local markets were being over-supplied, our growers made no attempt to supply outside markets. Now this is being done, and better means of handling and packing the fruit, so as to enable it to be shipped long distances, are now coming into vogue. With improved methods of handling and packing, we have a greatly extended market, in which we will have no local competition, hence will be able to secure good returns, so much so that I consider that grape-growing in Queensland has a very promising outlook for some years to come at any rate. In addition to growing grapes to supply the fresh-fruit trade and for winemaking, our western country is capable of producing good raisins and sultanas. So far, this industry has not been entered into commercially, the fresh fruit realising far too high a price for it to pay to convert it into raisins. Still, with increased production, this will have to take place, and when it does I am of opinion that we will be able to turn out a very saleable article. The growing of grapes here certainly requires considerable experience of a practical nature. This is not at all hard to obtain, and there are no insurmountable difficulties to the beginner, once he has learnt how to work his land so as to cause it to retain moisture during a dry spell, and to plant and prune his vines. These are matters in which any beginner can obtain practical advice from the Queensland Agricultural Department, as the Government of Queensland, recognising the importance of fruit-growing, grape-growing, and general agriculture to the State, have devoted considerable sums of money to the establishment of experiment farms, orchards, and vineyards in different parts of the State. All these Government institutions are under the control of thoroughly qualified managers, who are willing at all times to give any assistance to beginners, thereby enabling the latter to keep free from mistakes, and to obtain the best returns as the result of their labour. Instructors, thoroughly conversant with the State as a whole, are also available for giving practical advice, so that there is no necessity for a beginner, through lack of experience, to waste any time in finding out for himself what his soil and climate are suited for. He can start on the right lines from the beginning, and keep to right lines if he will only take advantage of the advice, based on practical experience, that is given him. Queensland is a good land for the intending fruit-grower. We offer you good soil, a choice of climates, suitable for the growing of practically every kind of commercial fruit, a healthy climate to live in, cheap land, free education for your children, and free advice from competent experts for yourselves. This is a country that has not been advertised or puffed up; that is, in consequence, not by any means well known; but it is a country that, taken all in all, will take a lot of beating when one is looking out for a home. Its natural advantages and the other inducements it offers to intending settlers, particularly those interested in fruit culture, cannot, in my opinion, be equalled, and certainly not excelled, elsewhere; and, as I stated in the beginning of this paper, my opinion is based on practical experience gained in various parts of the fruit-producing parts of the world.

Black Mammoth Grape.   Cinsaut Grape.

List of Fruits Grown in Queensland.

Almonds, several varieties
Almond, Fiji
Apples, many varieties
Apricots, many varieties
Averrhoa
Avocada Pear
Bael Fruit
Banana, several varieties
Barberry
Blackberry
Brazilian Cherry
Bread Fruit
Burdekin Plum
Carob Bean
Chalta
Cherries, several varieties
Chestnut — Spanish
Chestnut — Japanese
Chinese Raisin
Citrons, several varieties
Cocoa-nut, many varieties
Custard Apples (Cherimoyers)
Dates
Davidsonia Plum
Figs, several varieties
Gooseberries — Cape
Gooseberries — Otaheitan
Granadillas
Grapes, many varieties
Guavas, many varieties
Jujube
Kai Apple
Kumquat
Litchi
Longan
Lemons, several varieties
Limes, several varieties
Loquats
Mandarins, several varieties
Mangoes, many varieties
Mangosteen — Sour or Coochin York
Medlars
Melons, many varieties
Monstera
Mulberries, several varieties
Natal Plum
Nectarines, several varieties
Olives, several varieties
Oranges, many varieties
Papaw, several types
Passion Fruit, several types
Peaches — China, several varieties
Peaches — Ceylon, several varieties
Pears, many varieties
Pecan Nut
Persimmons, several varieties
Pineapples, several varieties
Pistachio Nut
Plums — European, several varieties
Plums — Japanese, several varieties
Plums — American, several varieties
Pomegranate
Quince — European, several varieties
Quince — Japanese
Queensland Nut
Raspberries, several types
Rosellas
Rose Apple
Sapodilla Plum
Shaddock or Pomelo, several types
Star Apple
Strawberries, many varieties
Tamarinds
Tree Tomato
Vi Apple
Walnut

Peaches—Persian, many varieties   Whampee

List of Vegetables Grown in Queensland.

- Artichokes — Jerusalem and Globe
- Asparagus
- Beans of all kinds
- Beetroot
- Broccoli
- Brussels Sprouts
- Cabbage
- Cabbage — Chinese
- Capsicums
- Cardoons
- Carrots
- Cassava
- Cauliflowers
- Celery
- Chicory
- Chokos
- Cress
- Cucumbers
- Earth Nuts (Peanuts)
- Egg Plant
- Endive
- Eschalots
- Garlic
- Herbs — all kinds
- Horseradish
- Kohl-rabi
- Leeks
- Lettuce
- Mushrooms
- Mustard
- Nasturtiums
- Ockra
- Onions
- Peas
- Potatoes — English and Sweet
- Pumpkins
- Radishes
- Rhubarb
- Salsify
- Seakale
- Spinach
- Squashes
- Sweet Corn
- Swedes
- Taro
- Tomatoes
- Turnips
- Vegetable Marrows
- Yams

www.ingramcontent.com/pod-product-compliance
Lightning Source LLC
Chambersburg PA
CBHW031434210526
45464CB00005B/2192